Perturbation Methods

Perturbation Methods

P. Kumar

 **BS Publications**

An Imprint of **BSP Books Pvt. Ltd.**
4-4-309/316, Giriraj Lane,
Sultan Bazar, Hyderabad - 500 095.

Published by:

 **BS Publications**

An Imprint of **BSP Books Pvt. Ltd.**
4-4-309/316, Giriraj Lane, Sultan Bazar,
Hyderabad - 500 095.
Phone: 040 -23445688
e-mail: info@bspbooks.net
website: www.bspbooks.net

ISBN: 978-93-95038-83-6 (Paperback)

To my brilliant daughter Dr. Urvashi Shrivastav, whose pursuit of knowledge led her to earn a PhD, and to my exceptional son-in-law Vineet Beri, a cybersecurity expert whose skills protect us in the digital age. Your dedication to learning and your commitment to securing our digital world inspire me daily. This book is dedicated to both of you, as a testament to your intelligence, dedication, and the love you bring to our family. May your future be filled with continued success, happiness, and the joy of making the world a safer place.

With all my love, Dr. Prabhat Kumar

Preface

In the vast realm of science, there exists a fascinating approach that allows us to unravel the secrets of complex systems by peering into their subtle disturbances. This method, known as perturbation theory, acts as a lighthouse for comprehending the complex dance of particles, forces, and interactions that governs the universe. We explore the tremendous ramifications of perturbation theory across several scientific areas as we travel through the pages of this book.

I would like to express my heartfelt gratitude to the Almighty; and to all those who have contributed to the creation of this book. To the mentors and educators whose wisdom and guidance have shaped my understanding of perturbation theory, thank you for your dedication to imparting knowledge. To the researchers and scientists who have paved the way with their ground breaking work, thank you for inspiring this exploration.

To my friends and colleagues who provided invaluable discussions and insights, thank you for your camaraderie and support. To my family, whose unwavering encouragement and belief in me have been my driving force, thank you for always being my foundation. Special thanks to Poonam Rane and Shradha Chhajed for their valuable contribution in reviewing the manuscript of the book. Finally, to the readers who embark on this intellectual journey, thank you for your curiosity and open-mindedness.

Author

Contents

About the Author

Dr. P. Kumar is an eminent mathematician and meteorologist with over 5 decades of rich and varied experience across multiple disciplines.

Born in 1948. He completed his graduation from Gorakhpur University, India and his post-graduation from Banaras Hindu University, India. He completed his PhD in mathematics from Banaras Hindu University in 1973. His PhD thesis was comprised of six research papers. He then proceeded to Post Doctoral Research Fellowship with CSIR in artificial rain making. He received "President of India Medal" for meritorious services.

He was commissioned the Indian Air Force (IAF) as an officer in the Meteorological Branch. He served for 22 years in various locations such as Gwalior, Hyderabad, Kumbigram, Ambala, and Coimbatore and retired as Wing Commander. He performed various important duties such as Operational Weather Forecaster, Meteorological Instructor & Research Guide, Research Scientist, PhD guide, Administrator etc.

Having always nurtured a keen interest in academics, he entered the field after retirement and had various successful stints across India. He has taught in various prestigious institutions such as National Law University Jodhpur, Aravali Institute of Management Jodhpur etc. He retired as Head of Department (Core Engineering and Engineering Sciences), MIT College of Engineering, Pune.

He has a rich and varied career in research. He has conducted research in prestigious institutions such as DRDO and CSIR etc.

He has authored eight books on Mathematics, Meteorology, Hydraulics, Physics, Hailstorm Prediction and control, hydraulic machines, severe storms etc.

He has published more than 60 papers in various international and national journals on varied topics. Areas of his research papers are as follows:

1. Theoretical Fluid Dynamics
2. Magneto Hydrodynamics
3. Meteorological observations
4. Forecasting
5. Theoretical Meteorology

His theory of tornado genesis is the first and only theory of its kind in the world.

In his illustrious career, he has received many awards. **He is the recipient of American Meteorological Society's "Award for Outstanding Contribution to the Advance of Applied Meteorology" in 2022 and is the only scientist outside of USA to be recognized with this award.** He won the Best Paper Presentation Award at the International Conference on Meteorology and Climate Sciences held at University of Dhaka, Bangladesh.

He has performed the role of Guest /Keynote speaker in many national and international conferences and seminars. Key engagements include Taiwan University of Taipei, National Central University Taiwan, Indian Agricultural Research University, Air Force Administrative College, Indian Institute of Science Education and Research Mohali, Punjab Agriculture University.

He is a Life Member of Indian Meteorological Society and Indian Mathematical Society.

He lives in Pune, Maharashtra India.

Chapter 1

Origin

Perturbation theory comprises mathematical methods for finding an approximate solution to a problem, by starting from the exact solution of a related, simpler problem. A critical feature of the technique is a middle step that breaks the problem into "solvable" and "perturbation" parts. Perturbation theory is applicable if the problem at hand cannot be solved exactly, but can be formulated by adding a "small" term to the mathematical description of the exactly solvable problem.

Perturbation theory leads to an expression for the desired solution in terms of a formal power series in some "small" parameter - known as a **perturbation series** - that quantifies the deviation from the exactly solvable problem. The leading term in this power series is the solution of the exactly solvable problem, while further terms describe the deviation in the solution, due to the deviation from the initial problem. Formally, we have for the approximation to the full solution A, a series in the small parameter.

If A_0 would be the known solution to the exactly solvable problem and $A_1, A_2, \ldots$ represents the **higher-order terms** which may be found iteratively by some systematic procedure, for small ϵ these higher-order terms in the series becomes successively smaller. Furthermore, if the small parameter is zero, then the given equation is exactly solvable, and the problem is reduced to finding the asymptotic behavior of the best approximation to the true solution.

An approximate "perturbation solution" is obtained by truncating the series, usually by keeping only the first two terms, the initial solution and the "first order" perturbation correction. If truncated after first term then it is termed as "second order" perturbation correction.[1]

1.1 Historical Development

Initially, Perturbation theory was proposed for the solution of problems in celestial mechanics, particularly to study the motions of planets in the solar system, As the mass of each planet is very less as compared to that of the Sun and the planets are very remote from each other, the gravitational forces between the planets can be neglected, and the planetary motion is taken into consideration to the first approximation as taken along Kepler's orbits (Appendix - A 1.7.1), which are defined by the equations of the two-body problem (Appendix - A 1.7.2), the two bodies being the planet and the Sun. The solution to the two-body problem can be obtained using Kepler's law which suggests that the point mass planet moves on an elliptical path with the Sun at one of the two foci.

The dynamics become extremely complicated when gravitational influence from any other body is added. Since the available astronomical data is of high accuracy, considering how the motion of a planet around the Sun is affected by other planets becomes necessary. This was the origin of the three-body problem (Appendix - B 1.8.1); thus, in studying the system Moon-Earth-Sun the mass ratio between the Moon and the Earth was chosen as the small parameter. Later Poincare[2] showed that the three-body problem does not admit a sufficient number of prime integrals which allow to integration problem. This gave rise to the concept of restricted three body problem.

J. L. Lagrange (25 January 1736 - 10 April 1813), and P. Laplace (23 March 1749 - 05 March 1827) were the first ones to progress over the view that the constants describing the motion of the planet around the sun are perturbed, as they were due to the motion of the other planets. The constants vary as a function of time, hence the name Perturbation theory. Perturbation theory focuses on finding an approximate solution of nearly- integrable systems, i. e., systems which consists of an integrable part and a small perturbation. The core aspect of this theory is to construct a canonical transformation which eliminates the perturbation of higher orders. A typical example of a nearly-integrable system is provide by two-body problem greatly stimulated the development of perturbation theories. The dynamics of the solar system has always been a testing ground for such theories, whose applications extend from the computation of the ephemerides of the natural bodies to developing the trajectories for artificial satellites.

The dynamics of asteroids are driven by the Sun and perturbation by Jupiter, as the Jupiter-Sun mass ratio is approximately 10^{-3}. The solutions to this type of problem motivated the work of scientists around the XVIII and XIX centuries. Indeed, Laverrier (11 March 1811 - 23 September 1877), Delaunary (12 March 1890 - 17 July 1980), Tisserand (13 January 1845 - 20 October 1896) and Poincare (29 April 1854 - 17 July 1912) added to further develop perturbation theories which are at the basis of the study of the dynamics of celestial bodies.

1.2   Discovery of Neptune Planet

Neptune was the first planet to be discovered by using mathematics. Figure 1.1 is the pictorial representation of the solar system. After the discovery of Uranus in 1781, astronomers noticed that the planet was being pulled slightly out of its normal orbit. John Couch Adams of Britain and Urbain Jean Joseph Leverrir of France used mathematics to predict that the gravity from another planet beyond Uranus was affecting the orbit of Uranus. They figured out not only where the planet was, but also how much mass it had. Following the suggestion provided by the theoretical investigation a young astronomer, Johann Gottfried Galle decided to search for the predicted planet and observed Neptune for the first time on 23 September 1846. This discovery represented the first triumph of perturbation theory.

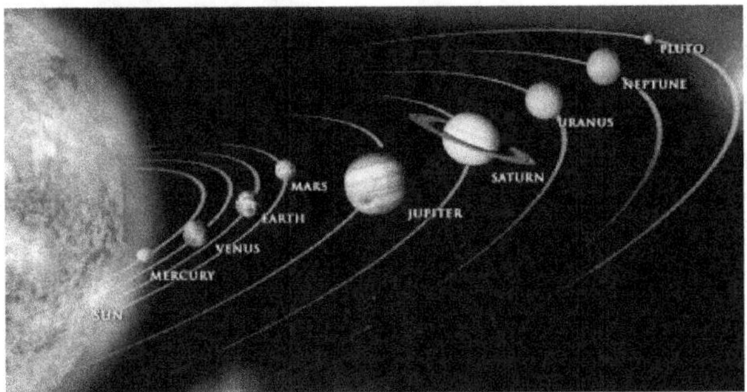

Figure 1.1: Solar System

Similar to the discovery of Neptune, Pluto was discovered in 1930 by

Clyde Tombaugh. Astronomers noticed that the orbits of Neptune and Uranus were being affected by the gravity of an unknown object in the solar system. Clyde Tombaugh carefully studied images of the night sky, and after a lot of hard work, he finally discovered Pluto. Interestingly Clyde Tombaugh was only 24 years old when he made this discovery.

Well-developed perturbation methods were adopted and adapted to solve new problems arising during the development of quantum mechanics in $20^{(th)}$-century atomic and subatomic physics. Paul Dirac developed perturbation theory in 1927 to evaluate when a particle would be emitted in radioactive elements. It was later named Fermi's golden rule.

1.3   Mixed term theory

The mathematical challenge involved initially in the development of the theory was that the term in the expansion consisted of the time parameter t independent of the sine and cosine functions. The contribution of such terms to the series is significant only when t is very large(of the order of several hundred years), even then only the first approximation is obtained instead of the accurate planetary motion. The secular terms i.e., the terms of the form At^n appears because the frequency of the motion(rotation) of the planet under study depends on the respective frequencies of other planets. If this kind of relation is allowed then both secular and mixed terms $Bt\cos(\omega t + \psi)$ appear in the solution. Thus, in the framework of perturbation theory the relation

$$\omega = \omega_0 + \epsilon\,\omega_1 \tag{1.1}$$

permits the following expansion with respect to $\epsilon(\epsilon \ll 1)$:

$$\sin\omega t = \sin\omega_0 t + \epsilon\,\omega_1\,t\cos\omega_0 t + \ldots \tag{1.2}$$

The mixed term in this equation is obtained by expanding oscillations of frequency (1.1) by oscillations with frequency ω_0 .

Lindstedt (27 June 1854 - 16 May 1939), Paul Guldin (1577 - 1643), Ch. Delaunay (1890-1980),B. Bohlin, and S. Newcomb's (March 12, 1835 - July 11, 1909) work led to the development of special methods in Perturbation theory, which eliminates the secular terms and hence, it permits one to obtain a purely trigonometric solution. The expansion of the frequencies affected by the secular terms is no more an expansion with respect to the small parameter which means such expansion does not contain zero approximation frequencies instead it contains the frequencies which to some extent have been redefined or in modern physical terms, it is renormalized. Therefore, in perturbation theory, every member of the power series consisting of powers of a small parameter is convergent.

1.4   Small denominator problem

The method for the construction of a special kind of periodic solution was proposed by H. Poincare and A.M. Lyapunov and this method is efficient not only in the problems related to celestial mechanics but also used in the theory of differential equations in general.

The order of perturbation can be reduced using the method of successive canonical change of variables. This method also allows to take advantage of better convergence(superconvergence) in order to overcome the divergence of the series which appears due to the presence of small denominators in each series no matter what order, with the help of suitable canonical transformation.

1.5   Theory of Oscillations

Subsequent advances in perturbation theory are connected with the development of the theory of oscillations, especially with the development of the theory of non-linear oscillations. Non-linear ordinary differential equation of Rayleigh (1.3)

$$\ddot{x} + F(\dot{x}) + x = 0, \qquad \dot{x} = \frac{dx}{dt} \qquad (1.3)$$

where $F(u)$ satisfies the assumption $uF(u) < 0$ for small (u) and $uF(u) > 0$ for large (u) desirable auto-oscillation with one degree of freedom. The special case of Rayleigh equation is when

$$F(u) = -\mu(u - \frac{u^2}{3}), \qquad \mu = \text{constant}$$

It is known as The van der Pol equation, a special case of equation (1.3) in order to solve the equation

$$\ddot{x} - \mu(1 - x^2)\dot{x} + x = 0, \qquad \mu = const > 0, \qquad \dot{x}(t) \equiv \frac{dx}{dt} \qquad (1.4)$$

Van der Pol proposed the method of slowly varying coefficients that resembled the method used by Lagrange in solving problems of celestial mechanics. This method of slowly varying coefficients comprises of representation of the equation in which functions harmonically oscillate and their amplitudes and phases slowly vary as the function of the parameter t.

N. M. Krylov and N. N. Bogolyubov developed general theory of non-linear oscillations and used asymptotic methods in perturbation theory. Consider equation

$$\frac{d^2x}{dt^2} + \omega^2 x = \epsilon f\left(x, \frac{dx}{dt}, \epsilon\right) \tag{1.5}$$

In the case $\epsilon = 0$, equation 1.5 describes the oscillations that are purely harmonic which have a constant amplitude and a uniformly recurring phase. If $\epsilon \neq 0$, i.e. in the presence of a non-linear perturbation solution of equation 1.5 involves overtones, the dependence of instantaneous frequency on the amplitude, and a systematic increase/ decrease of the amplitude depending on the input or output of energies due to perturbing forces.

$$x = a\cos\psi + \epsilon u_1(a, \psi) + \epsilon^2 u_2(a, \psi) + \ldots \tag{1.6}$$

where $u_i(a, \psi), i = 1, 2, \ldots$ represent periodic functions of angle ψ with period 2π, The values of a and ψ are taken as functions of time and are defined by the differential equations.

$$\left.\begin{array}{l} \frac{da}{dt} = \dot{a} = \epsilon A_1(a) + \epsilon^2 A_2(a) + \ldots \\ \frac{d\psi}{dt} = \dot{\psi} = \omega + \epsilon B_1(a) + \epsilon^2 B_2(a) + \ldots \end{array}\right\} \tag{1.7}$$

Thus, the problem is reduced to the choice of suitable expressions for the functions $u_i(a, \psi), A_i(a), B_i(a), i = 1, 2, \ldots$ so that the expression (1.6),along with a and ψ after replacing by functions dependent on time as in equation 1.7, becomes the solution of the original equation 1.5. Additional conditions are imposed to ensure that the secular terms (terms of the form At^n) do not appear in the solution.

If in the formal series 1.6 the expansion is cut-off after $(m+1)^{th}$ terms, one obtains the m-th approximation; this approximation is asymptotic i.e. if m is fixed and *epsilon* $\to 0$, then the expression 1.6 approximates the exact solution of 1.5. Van der Pol equation and the first approximation equations are identical.

1.6 Averaging Method

Differential equations describing oscillatory processes and containing a "small" parameter may often be reduced to standard form

$$\frac{dX_s}{dt} = \epsilon\, X_s(t, x_1, \ldots), x_n), \ s = 1, \ldots, n \tag{1.8}$$

where ϵ is a small positive parameter. Several problems in physics and technology can be reduced to this form. Averaging method, a special method of approximation is developed for systems of differential equations as in 1.8. This method suggests that if values of ϵ are sufficiently small on a finite interval, the change of variables can be used to obtain the average equations.

$$x_i = \xi_i + \epsilon X_i$$

can be used to obtain the averaged equations

$$\frac{d\xi_s}{dt} = \epsilon\, X_s, 0(\xi_1, \ldots, \xi_n), \ s = 1, \ldots, n \tag{1.9}$$

where

$$X_s, 0(\xi_1, \ldots, \xi_n) = \lim_{T \to \infty} \frac{1}{T} \int_0^T X_s(t, \xi_1, \ldots, \xi_n)\, dt$$

The number of criteria for the existence and stability of auto-oscillatory systems can also be obtained by the averaging method.

Equation 1.9 give the estimates of the difference $|X_i - \epsilon_i|$ over a time interval of length L/ϵ.

1.7 Appendix - A

1.7.1 The approximate nature of Kepler's laws and two body Problem

The constraints placed on the <u>force</u> for Kepler's laws to be derivable from Newton's laws were that the force must be directed toward a central fixed point and that the force must decrease as the inverse square of the distance. In actuality, however, the sun. which serves as the source of the major force, is not fixed but experiences small accelerations because of the planets, in accordance with Newton's second and third laws. Furthermore, the planets attract one another, so that the total force on a planet is not just that due to the sun, other planets <u>perturb</u> the elliptical <u>motion</u> that would have occurred for a particular planet if that planet had been the only orbiting an isolated <u>Sun</u>. Keplar's laws, therefore, are only approximate. The motion of the Sun itself means that, even when the attractions by other planets are neglected, Kepler's third law must be replaced by $(M + m_i)\,\tau^2 \propto a^3$, where m_i is one of the planetary masses and M is the Sun's mass. The Kepler's law are such good approximations to the actual planetary results from the fact that all the planetary masses are very small compared to that of the Sun. The perturbations of the elliptic motion are therefore small, and the coefficient $M + m_i \equiv M$ for all the planetary masses m_i means that Kepler's third law is very close to being true.

Newton's second law for a particular mass is a second-order <u>differential equation</u> that must be solved for whatever forces may act on the body if its position as a function of time is to be deduced. The exact solution of this equation, which resulted in a derived trajectory that was an ellipse, parabola, or hyperbola, depended on the assumption that there were only two point particles by the inverse square force. Hence, this "<u>gravitational two-body problem</u>" has exact Solution that reproduces Kepler's laws. If one or more additional bodies also interact with the original pair through their mutual gravitational interactions, no exact solution for differential equations of motion of any of the bodies involved can be obtained. As was noted above, however, the motion of a planet is almost elliptical, since all masses involved are small compared to the Sun. it is then convenient to treat the motion of a particular planet as slightly perturbed elliptical motion and to determine the changes in the <u>parameters</u> of the ellipse that result from the small forces as time progresses. It is the elaborate developments of various <u>perturbation</u> theories and their applications to approximate the exact motion of celestial bodies that has occupied celestial mechanicians since Newton's time.

1.7.2   Two-body problem

A problem dealing with the motion of two material points P_1 and P_2 with masses m_1 and m_2 respectively, moving in three-dimensional Euclidean space E^3 when acted upon by the mutual Newton attracting forces. The problem is special case of the n-body problem, which may be described by a system of ordinary differential equations of order $6n$, and has 10 independent integrals: 6 of motion of the centre of inertia, 3 of law of areas (equivalently, conservation of angular momentum) and 1 of energy conservation. The two-body problem also has three Laplace integrals (one of which is independent of the preceding ones) and is completely integrable.

The integration of the two-body problem is more conveniently effected in a special system of coordinates, in which these integrals are employed. If the origin of the Cartesian coordinates x, y, z is placed at the centre of masses $(m_1 r_1 + m_2 r_2)/(m_1 + m_2)$ and the axis z is directed along the relative angular momentum vector, then the motion of the relative position vector $r_1 - r_2 = (x, y, z)$ takes place in the plane $z = 0$ and satisfies the system

$$\mu\ddot{x} = -fxr^{-3}, \qquad\qquad \mu\ddot{y} = -fyr^{-3}, \qquad\qquad (1)$$

where $r = \sqrt{x^2 + y^2}$, $\quad \mu = m_1 m_2/(m_1 + m_2)$ is the reduced mass and f is the gravitational constant. The system (1) has four integrals:

$x\dot{y} - y\dot{x} = x$ (law of areas),

$$\frac{1}{2}\mu(\dot{x}^2 + \dot{y}^2) - fr^{-1} = h \text{ (energy)}$$

$\mu^2 c\dot{y} - \mu fxr^{-1} = \lambda_1$ and $\mu^2 c\dot{x} + \mu fyr^{-1} = \lambda_2$ (Laplace),

which are interconnected by the relation

$\lambda_1^2 + \lambda_2^2 = 2\mu^3 hc^2 + \mu^2 f^2$

here

$$c^2 = \lambda_1 x + \lambda_2 y + \mu r, \qquad\qquad (2)$$

i.e. the orbits of the relative position vector are conical sections with parameter $p = c^2/\mu$, major semi-axis $a = -\mu/(2h)$, eccentricity $e = \mu^{-1}\sqrt{1 + 2hc^2}$, longitude of pericentre $\omega(\lambda_1 = \mu e\cos\omega, \lambda_2 = \mu e\sin\omega)$, and with the focus at the coordinate origin. the location of the relative positive vector on the orbit is determined by the true anomaly v, counted from the direction towards the pericentre; (2) then implies that $r = p/(1 + e\cos v)$. if $c \neq 0$, three types of orbits are possible:

1. If $h < 0$, they are ellipses.

2. if $h > 0$, they are hyperbolas.

3. If $h = 0$, they are parabolas.

If $c = 0$, the motion is rectilinear. The two-body problem describes an unperturbed Kepler motion of a planet with respect to the Sun or of a satellite respect to a planet, etc.

1.8 Appendix B

1.8.1 Three-body problem

The problem on the motion of three bodies, regarded as material points, mutually attracting one another according to Newton's law of gravitation (cf. <u>Newton laws of mechanics</u>). The classical example of the three-body problem is that of the motion of the Sun-Earth-Moon system. The three-body problem consists in finding the general solution of the system of differential equations

$$m_i \frac{d^2 x_i}{\partial t^2} = \frac{\partial U}{\partial x_i}, \qquad m_i \frac{d^2 y_i}{\partial t^2} = \frac{\partial U}{\partial y_i}, \qquad m_i \frac{d^2 z_i}{dt^2} = \frac{\partial U}{\partial z_i}, \qquad i = 1, 2, 3,$$

where x_i, y_i, z_i are the rectangular coordinates of the body M_i in some absolute coordinate frame with fixed axes, t is the time, m_i is the mass of M_i, and U is the potential, which depends only on the mutual distances between the points. The function U is defined by the relation

$$U = f \left(\frac{m_1 m_2}{\Delta_{12}} + \frac{m_2 m_3}{\Delta_{23}} + \frac{m_3 m_2}{\Delta_{13}} \right), f > 0,$$

where the mutual distances $\Delta_{ij}, i, j = 1, 2, 3$ are given by the formula

$$\Delta_{ij} = \Delta_{ji} = \sqrt{(x_i - x_j)^2 + (y_i - y_j)^2 + (z_i - z_j)^2}$$

From the properties of the potential one can derive ten first integrals of the equations of motion in the absolute system of coordinates, Six of them, called the integrals of motion of the centre of mass, determine the uniform rectangular motion of the centre of mass of the three bodies. The three integrals of the angular momentum fix the value and the direction of the angular momentum of the three body system. The energy integral defines the constant magnitude of total energy of the system. H. Bruns (1887) proved that the equations of motion of the three-body problem have no other first integrals expressible in terms of algebraic functions of the coordinate and their derivatives. H. Poincare (1889) further proved that the equations of motion of the three-body problem do not have transcendental integrals expressible in terms of single-values analytic functions. C. Sundman (1912) found the general solution of the problem in the form of power series in a certain regularizing variable, converging at each instant. However, the Sundman series proved to be completely useless for qualitative investigations as well as for practical computations due to its extremely slow convergence.

The equation of three-body problem admits five particular solutions, in which all three material points are in some fixed plane. Here, the configuration of the three bodies remain fixed and they describe Kepler trajectories with a common focus at the centre of mass of the system. Two of the particular solutions corresponds to the case when the three bodies form an equilateral triangle at all times. This is the so-called triangular solution of the three-body problem, or the Lagrange solutions corresponding to three bodies on one straight line are called the rectilinear solutions, or the Euler solutions.

For the general solution of the three-body problem, final motions have been studied in detail, that is, the limiting properties of the motion as $t \to +\infty$ and $t \to -\infty$.

A particular case of the three-body problem is the so-called restricted three-body problem, which is obtained from the general three-body problem in case the mass of one of the three bodies is so small that its influence on the motion of the other two bodies can be neglected. In this case, the bodies M_1 and M_2 with finite masses m_1 and m_2 move under the action of their mutual attraction along Kepler orbits. In the right-handed rectangular coordinate system $G\xi\eta\zeta$ with origin G at the centre of mass of M_1 and M_2, with axis ξ directed along the line joining M_1 and M_2 and axis ζ perpendicular to the plane of their motion, the motion of the third body M_3 of small mass is described by the following differential equation:

$$\ddot{\xi} - 2\dot{\nu}\ \dot{\eta} - \dot{\nu}^2\ \xi - \ddot{\nu}\ \eta = \frac{\partial W}{\partial \xi},$$

$$\ddot{\eta} - 2\dot{\nu}\ \dot{\xi} - \dot{\nu}^2\ \eta - \ddot{\nu}\ \xi = \frac{\partial W}{\partial \eta},$$

$$\ddot{\zeta} = \frac{\partial W}{\partial \zeta},$$

where $W = f\left(\dfrac{m_1}{r_1} + \dfrac{m_2}{r_2}\right)$,

ν is the true anomaly of the Kepler motion of M_1 and M_2, and r_1 and r_2 are the distances of M_3 from M_1 and M_2, respectively. In the case of the circular restricted three-body problem,

$$\dot{\nu} = n = const, \qquad \ddot{\nu} = 0,$$

the equation of motion of M_3 have also a first integral, called the jacobi integral, of the form

$$\dot{\xi}^2 + \dot{\eta}^2 + \dot{\xi}^2 = n^2(\xi^2 + \eta^2) + 2f\left(\frac{m_1}{r_1} + \frac{m_2}{r_2}\right) + C$$

where C is an arbitrary constant. The surface defined by the equation

$$n^2(\xi^2 + \eta^2)2f\left(\frac{m_1}{r_1} + \frac{m_2}{r_2}\right) + C = 0$$

is called the surface of zero velocity and is remarkable in that it determines the regions of possible motions of M_3 relative to M_1 and M_2. The restricted three-body problem has particular solutions similar to those of the general three-body problem. The position of the body with a small mass in these particular solutions is called the points of libration.

For the restricted three-body problem, various classes of periodic motion have been investigated.

Chapter 2

Regular perturbation theory

2.1 Taylor expansion and Perturbation Theory

A clear understanding of Taylor expansion is the primary step for the regular perturbation theory.

2.1.1 Taylor's Theorem (Generalized Mean Value Theorem)

Let $f(x)$ be defined in $[a, a+h]$ such that $f(x), f'(x), f''(x), \ldots f^{(n-1)}(x)$ be continuous in $[a, a+h]$ and $f^{n}(x)$ exist for $\forall x \in (a, a+h)$ then there exist at least one number $\theta(0 < \theta < 1)$ such that

$$f(a+h) = f(a) + hf'(a) + \frac{h^2}{2}f''(a) + \cdots + \frac{h^{n-1}}{(n-1)}f^{(n-1)}(a) + R_n$$

where R_n is the reminder after n terms and is given by

1. Schlomilch and Roche form of remainder:

$$R_n = \frac{h^n(1-\theta)^{n-p}}{(n-1)!P}f^{(n)}(a+\theta h) \tag{2.1}$$

where $h = b - a, 0 < \theta < 1$ and P is a positive rational number.

2. Cauchy's form of remainder: (Put $P = 1$ in (2.1))

$$R_n = \frac{h^n(1-\theta)^{n-1}}{(n-1)!} f^{(n)}(a+\theta h)$$

3. Lagrange's form of remainder: (Put $P = n$ in (2.1))

$$R_n = \frac{h^n}{n!} f^{(n)}(a+\theta h)$$

2.1.2 Taylor's Series

If the remainder $R_n \to 0$ as $n \to \infty$ then Taylor's theorem assumes the form of Taylor's series and it is

$$f(a+h) = f(a) + hf'(a) + \frac{h^2}{2!} f''(a) + \cdots + \frac{h^n}{n!} f^{(n)}(a) + \ldots$$

Different forms of Taylor's series are as follows.

(i) The first form is

$$f(x+h) = f(x) + hf'(x) + \frac{h^2}{2!} f''(x) + \cdots + \frac{h^n}{n!} f^{(n)}(x) + \ldots$$

Here the expansion is in ascending power of h.

(ii) Interchanging x and h in the above form, we get

$$f(x+h) = f(h) + xf'(h) + \frac{x^2}{2!} f''(h) + \cdots + \frac{x^n}{n!} f^{(n)}(h) + \ldots$$

Here the expansion is in ascending power of x.

(iii) Putting $h = x - a$ in Taylor's series we get

$$f(x) = f(a)+(x-a)f'(a)+\frac{(x-a)^2}{2!} f''(a)+\cdots+\frac{(x-a)^n}{n!} f^{(n)}(a)+\ldots$$

Here the expansion is in ascending power of $x - a$ or the expansion at $x = a$

2.1.3 Regular Perturbation Expansion

The main paradigm of Taylor's expansion is that a small change of ε to x makes a small change to $f(x)$. This is also the basis of regular perturbation expansions. Therefore, the basic principle and practice of regular perturbation expansion can be summarized as under:-

(i.) Set $\varepsilon = 0$ and solve the resulting system (solution f_0 for definiteness.)

(ii.) Perturb the system by allowing ε to be nonzero (but small)

(iii.) Formulate the solution to the new, perturbed system as a series

$$f_0 + \varepsilon f_1 + \varepsilon^2 f_2 + \cdots$$

Compare it from Taylor's Expansion assuming $f_1 = \frac{f'}{1!}$, $f_2 = \frac{f''}{2!}, \ldots$

(iv.) Expand the governing equations as a series in ε, collecting terms with equal powers of ε; solve them in term as far as the solution is required.

2.2 Perturbation Method for Solution of Algebraic Equation

2.2.1 Illustrative Example 1

Consider the quadratic equation,

$$X^2 - 1 = \varepsilon X \tag{2.2}$$

The two roots of this equation are

$$X_1 = \varepsilon/2 + \sqrt{1 + \varepsilon^2/4}, \qquad X_2 = \varepsilon/2 - \sqrt{1 + \varepsilon^2/4} \tag{2.3}$$

For small ε, these roots are well approximated by the first few terms of their Taylor series expansion (see figure)

$$X_1 = 1 + \varepsilon/2 + \varepsilon^2/8 + O(\varepsilon^3), \qquad X_2 = -1 + \varepsilon/2 - \varepsilon^2/8 + O(\varepsilon^3) \tag{2.4}$$

Can we obtain (2.4) without prior knowledge of the exact solution of (2.2)? Yes, using regular perturbation theory. The technique involves the following four steps.

STEP A. Assume that the solution(s) of (1.1) can be Taylor expanded in ε. Then we have

$$x = X_0 + \varepsilon X_1 + \varepsilon^2 X_2 + O(\varepsilon^3), \tag{2.5}$$

for X_0, X_1, X_2 to be determined.

STEP B. Substitute (2.5) into (2.2) written as $x^2 - 1 - \varepsilon x = 0$, and expand the left-hand side of the resulting equation in power series of ε.

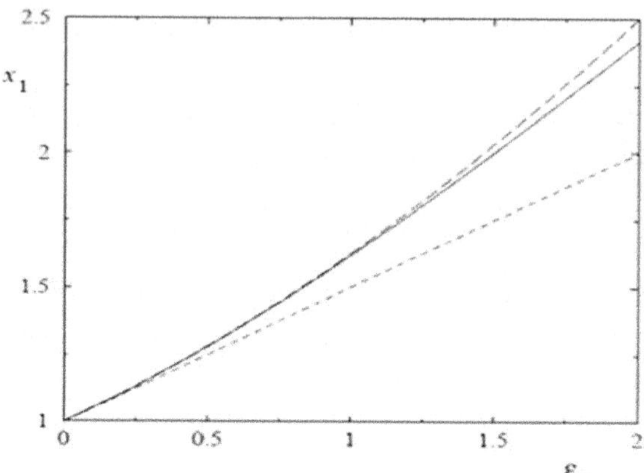

Figure 2.1: The root x_1 plotted as a function of ε (solid line), compared with the approximations by truncation of the Taylor series at $O(\varepsilon^2), x_1 = 1 + \varepsilon/2$ (dotted line), and $O(\varepsilon^3), x_1 = 1 + \varepsilon/2 + \varepsilon^2/8$ (dashed line). Notice that even though the approximations are *a priori* valid in the range of $\varepsilon \leq 1$ only, the approximation $x_1 = 1 + \varepsilon/2 + x^2/8$ is fairly good even up to $\varepsilon = 2$.

Using

$$x^2 = X_0^2 + 2\varepsilon X_0 X_1 + \varepsilon^2(X_1^2 + 2X_0 X_2) + O(\varepsilon^3), \qquad (2.6)$$

this gives

$$X_0^2 - 1 + \varepsilon(2X_0 X_1 - X_0) + \varepsilon^2(X^2 + 2X_0 X_2 - X1) + O(\varepsilon^3) = 0 \quad (2.7)$$

STEP C. Equate to zero the successive terms of the series in the left-hand side of (2.7):

$$
\begin{aligned}
O(\varepsilon^0): &\quad X_0^2 - 1 = 0, \\
O(\varepsilon^1): &\quad 2X_0 X_1 - X_0 = 0, \\
O(\varepsilon^2): &\quad X_1^2 + 2X_0 X_2 - X_1 = 0, \\
O(\varepsilon^3): &\quad \ldots
\end{aligned}
\qquad (2.8)
$$

STEP D. Successively solve the sequence of equations obtained in (2.8). Since $X_0^2 - 1 = 0$ has two roots, $X_0 = \pm 1$, one obtains

$$
\begin{aligned}
X_0 = 1, &\quad X_1 = 1/2, &\quad X_2 = 1/8 \\
X_0 = -1, &\quad X_1 = -1/2, &\quad X_2 = -1/8
\end{aligned}
\qquad (2.9)
$$

It can be checked that substituting (2.9) into (2.5) one recovers (2.4).

2.2.2 Illustrative Example 2

From the previous example, it might not be clear what the advantage of regular perturbation theory is, since one can obtain (2.4) more directly by Taylor expansion of the roots in (2.3). To see the strength of regular perturbation theory, consider the following equation

$$X^2 - 1 = \varepsilon e^x \tag{2.10}$$

The solutions of this equation are not available; therefore the direct

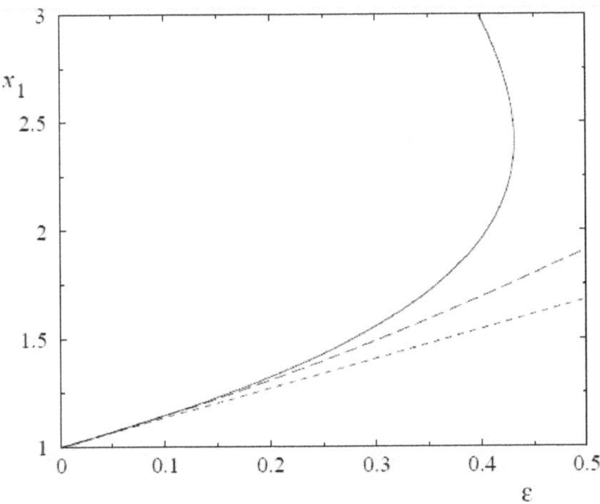

Figure 2.2: The solid line is the graph of two of the three solutions of (2.10) obtained numerically and plotted as a function of ε (solid line). Also plotted are the approximations by truncation of the Taylor series at $O(\varepsilon^2)$, $x_1 = 1 + \varepsilon e^2$ (dotted line), and $O(\varepsilon^3)$, $x_1 = 1 + \varepsilon e/2 + \varepsilon^2 e^2/8$ (dashed line).

method is inapplicable here. However, the Taylor series expansion of these solutions can be obtained by perturbation theory. We introduce the expansion (2.5) as in STEP A. In STEP B, we use (recall that $e^z = 1 + z + z^2/2 + O(z^3)$)

$$\begin{aligned}
\varepsilon e^x &= \varepsilon e^{X_0 + \varepsilon X_1 + \varepsilon^2 X_2 + O(\varepsilon^3)} \\
&= \varepsilon e^{X_0} e^{\varepsilon X_1 + \varepsilon^2 X_2 + O(\varepsilon^3)} \\
&= \varepsilon e^{X_0} + \varepsilon^2 X_1 e^{X_0} + O(\varepsilon^3)
\end{aligned} \tag{2.11}$$

Substituting this expression in (2.10) written as $x^2 - 1 - \varepsilon e^x = 0$ and using (2.6), we obtain

$$X_0^2 - 1 + \varepsilon(2X_0X_1 - e^{X_0}) + \varepsilon^2(X_1^2 + 2X_0X_2 - X_1e^{X_0}) + O(\varepsilon^3) = 0 \quad (2.12)$$

Thus, the sequence of equations obtained in STEP C is

$$
\begin{aligned}
O(\varepsilon^0): \quad & X_0^2 - 1 = 0, \\
O(\varepsilon^1): \quad & 2X_0X_1 - e^{X_0} = 0, \\
O(\varepsilon^2): \quad & X_1^2 + 2X_0X_2 - X_1e^{X_0} = 0, \\
O(\varepsilon^3): \quad & \qquad \cdots
\end{aligned}
\quad (2.13)
$$

from which we obtain (STEP D)

$$
\begin{aligned}
X_0 = 1, \quad & X_1 = e/2, \quad & X_2 = e^2/8 \\
X_0 = -1, \quad & X_1 = -1/(2e), \quad & X_2 = -1/(8e^2)
\end{aligned}
\quad (2.14)
$$

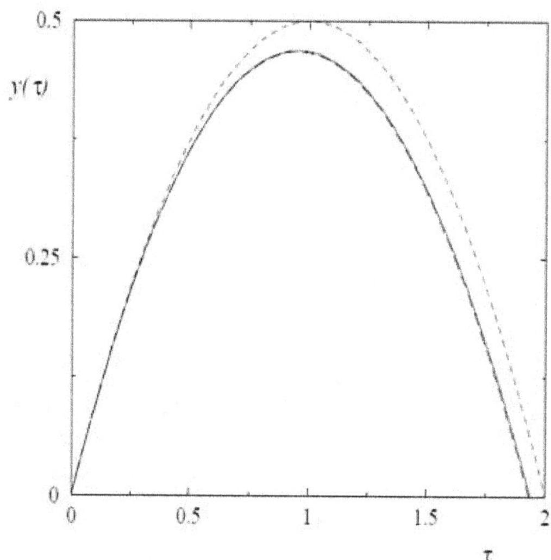

Figure 2.3: The exact solution (solid) line is compared with the approximations by truncation of the Taylor series at $O(\varepsilon)$ (dotted line), $O(\varepsilon^2)$ (dashed line), and $O(\varepsilon^3)$ (indistinguishable from the solid line).

or equivalently,

$$
\begin{aligned}
X_1 &= 1 + \varepsilon e/2 + \varepsilon^2 e^2/8 + O(\varepsilon^3) \\
X_2 &= -1 - \varepsilon/(2e) - \varepsilon^2/(8e^2) + O(\varepsilon^3)
\end{aligned}
\quad (2.15)
$$

The expression for x_1 is compared to the numerical solution of 2.10 on 2.2

Remark: In fact 2.10 has three solutions for $0 < \varepsilon < \varepsilon_1$ with $\varepsilon_1 \approx 0.43$. and only one for $\varepsilon > \varepsilon_1$. The solution which exists for all $\varepsilon > 0$ is the one with expansion given in X_2 in 2.15: the solution with the expansion given in X_1 in 2.15 disappears for $\varepsilon > \varepsilon_1$: the third solution (see figure 2.2: the solid line is the graph of a two-valued function) cannot be obtained by regular perturbation.
The expression for x_1 is compared to the numerical solution of 2.10 in figure 2.2.

Exercise 1: Solve by perturbation

$$x^2 - 4 = \varepsilon In(x) \tag{2.16}$$

Notice that, as $\varepsilon \to 0$, 2.16 reduces to the equation $x^2 - 4 = 0$, with two roots $x_{1,2} = \pm 2$. 2.16 also has two solutions (why?). How are they related to $x_{1,2} = \pm 2$? Can both solutions of 2.16 be obtained by perturbation?

2.3   Perturbation Theory of Differential Equation

2.3.1   Illustrative Example

Suppose we are trying to solve the following differential equation in $x \geq 0$:

$$\frac{df(x)}{dx} + f(x) - \varepsilon f^2(x) = 0, \qquad f(0) = 2 \tag{2.17}$$

Ignore the fact that we could have solved this equation directly! We'll use it as a model for more complex examples.
We first look at $\varepsilon = 0$

$$\frac{df(x)}{dx} + f(x) = 0, \qquad f(0) = 2 \Rightarrow f(x) = 2e^{-x}.$$

Now follow the perturbation method and set

$$f = 2e^{-x} + \varepsilon f_1(x) + \varepsilon^2 f_2(x) + \varepsilon^3 f_3(x) + \cdots$$

where in order to satisfy the initial condition $f(0) = 2$, we will have $f_1(0) = f_2(0) = f_3(0) = \cdots = 0$. Substituting into 2.17 gives

$$-2e^{-x} + \varepsilon f_1'(x) + \varepsilon^2 f_2'(x) + \varepsilon^3 f_3'(x) + 2e^{-x} + \varepsilon f_1(x)$$
$$+ \varepsilon^2 f_2(x) + \varepsilon^3 f_3(x) - 4\varepsilon e^{-2x} - 4\varepsilon^2 e^{-x} f_1(x)$$
$$- 4\varepsilon^3 e^{-x} f_2(x) - \varepsilon^3 f_1^2(x) = O(\varepsilon^4)$$

and we can collect powers of ε:

$$
\begin{aligned}
\varepsilon^0 &: -2e^{-x} + 2e^{-x} & = 0 \\
\varepsilon^1 &: f_1'(x) + f_1(x) - 4e^{-2x} & = 0 \\
\varepsilon^2 &: f_2'(x) + f_2(x) - 4e^{-x}f_1(x) & = 0 \\
\varepsilon^3 &: f_3'(x) + f_3(x) - f_1^2(x) - 4e^{-x}f_2(x) & = 0
\end{aligned}
$$

For the order ε^0 (or 1) equation is satisfied automatically. Now we simply solve for each order, applying the boundary condition as we go along.

Order ε terms.

$$
f_1'(x) + f_1(x) = 4e^{-2x} \Rightarrow f_1(x) = -4e^{-2x} + c_1 e^{-x}
$$

and the boundary condition $f_1(0) = 0$ gives $c_1 = 4$:

$$
f_1(x) = 4(e^{-x} - e^{-2x}).
$$

Order ε^2 terms.

The equation becomes

$$
f_2'(x) + f_2(x) = 4e^{-x}f_1(x) \Rightarrow f_2'(x) + f_2(x) = 16e^{-x}(e^{-x} - e^{-2x})
$$

with solution

$$
f_2(x) = 8(-2e^{-2x} + e^{-3x}) + c_2 e^{-x}
$$

and the boundary condition $f_2(0) = 0$ gives $c_2 = 8$:

$$
f_2(x) = 8(e^{-x} - 2e^{-2x} + e^{-3x}).
$$

Order ε^3 terms.

The equation is $f_3'(x) + f_3(x) - f_1^2(x) - 4e^{-x}f_2(x) = 0$ which becomes

$$
f_3'(x) + f_3(x) = 48(e^{-2x} - 2e^{-3x} + e^{-4x})
$$

The solution to this equation is

$$
f_3(x) = 16(-3e^{-2x} + 3e^{-3x} - e^{-4x}) + c_3 e^{-x}
$$

Applying the boundary condition $f_3(0) = 0$ gives $c_3 = 16$ so

$$
f_3(x) = 16(e^{-x} - 3e^{-2x} + 3e^{-3x} - e^{-4x})
$$

The solution we have found is:

$$f(x) = 2e^{-x} = 4\varepsilon(e^{-x} - e^{-2x}) + 8\varepsilon^2(e^{-x} - 2e^{-2x} + e^{-3x})$$
$$+ 16\varepsilon^3(e^{-x} - 3e^{-2x} + 3e^{-3x} - e^{-4x}) + \ldots$$

This is an example of a case where carrying out a perturbation expansion gives us an insight into the full solution. Notice that, for the terms we have calculated,

$$f_n(x) = 2^{n+1}e^{-x}(1 - e^{-x})^n$$

suggesting a guessed full solution

$$\sum_{n=0}^{\infty} e^{-x}2^{n+1}e^{-x}(1 - e^{-x})^n = 2e^{-x}\sum_{n=0}^{\infty}[2\varepsilon(1 - e^{-}x)]^n$$
$$= \frac{2e^{-x}}{1 - 2\varepsilon(1 - e^{-x})}$$

Having guessed a solution, of course, verifying it is straightforward: this is indeed the correct solution to the ODE of equation 2.17

2.3.2   Projectile equation

We take another example of a differential equation and initial condition as

$$\frac{d^2y}{d\tau^2} + \varepsilon\frac{dy}{d\tau} + 1 = 0, \qquad y(0) = 0, \qquad \frac{dy}{d\tau}(0) - 1 = 0, \qquad (2.18)$$

This equation governs the dynamics of a projectile thrown vertically into the air if air friction is taken into account. Here $\varepsilon = kV/(mg)$, where V is the initial velocity of the projectile, m is its mass, k is the friction constant ($[k]=MT^{-1}$) and g is the acceleration due to gravity. In 2.17, the altitude is measured in units of $l_1 = V^2/g$, and the time in units of $\tau = V/g$.
The exact solution of 2.18 is

$$y(\tau) = \frac{(1 + \varepsilon)}{\varepsilon^2}(1 - e^{-\varepsilon\tau}) - \frac{\tau}{\varepsilon} \qquad (2.19)$$

For $0 < \varepsilon \ll 1$, which corresponds to a situation where air friction is small, we can approximate $y(\tau)$ by the first few terms of its Taylor series expansion in ε (see figure 2.3). Using $1 - e^{-z} = z - z^2/2 + Z^3/3! + O(z^4)$ we obtain

$$y(\tau) = \tau - \tau^2/2 + \varepsilon(-\tau^2/2 + \tau^3/6) + \varepsilon^2(\tau^3/6 - \tau^4/24) + O(\varepsilon^3) \quad (2.20)$$

This expansion is uniformly valid in τ in the range $0 < \tau \ll 1/\varepsilon$, which is the range of physical interest since the projectile hits the ground well before $1/\varepsilon$ if $\varepsilon \ll 1$.

We wish to obtain the expansion in 2.20 without prior knowledge of the exact solution 2.19, using regular perturbation theory. We proceed similarly as for algebraic equations.

STEP A: Introduce the expansion

$$y(\tau) = y_0(\tau) + \varepsilon y_1(\tau) + \varepsilon^2 y_2(\tau) + O(\varepsilon^3) \tag{2.21}$$

where $y_0(\tau), y_1(\tau), y_2(\tau)$ are functions of τ to be determined.

STEP B: Substituting 2.20 into 2.17 (differential equation and initial condition) written as

$$\frac{d^2 y}{d\tau^2} + \varepsilon \frac{dy}{d\tau} + 1 = 0, \qquad y(0) = 0, \qquad \frac{dy}{d\tau}(0) - 1 = 0,$$

and expand the resulting equation in the power series of ε. This gives

$$\frac{d^2 y_0}{d\tau^2} + 1 + \varepsilon \left(\frac{d^2 y_1}{d\tau^2} + \frac{dy_0}{d\tau} \right) + \varepsilon^2 \left(\frac{d^2 y_2}{d\tau^2} + \frac{dy_1}{d\tau} \right) + O(\varepsilon^3) = 0,$$

$$y_0(0) + \varepsilon y_1(0) + \varepsilon^2 y_2(0) + O(\varepsilon^3) = 0, \tag{2.22}$$

$$\frac{dy_0}{d\tau}(0) - 1 + \varepsilon \frac{dy_1}{d\tau}(0) + \varepsilon^2 \frac{dy_2}{d\tau}(0) + O(\varepsilon^3) = 0.$$

STEP C: Equate to zero the successive terms of the series

$$O(\varepsilon^0) : \frac{d^2 y_0}{d\tau^2} + 1 = 0, \qquad y_0(0) = 0, \qquad \frac{dy_0}{d\tau}(0) - 1 = 0$$

$$O(\varepsilon^1) : \frac{d^2 y_1}{d\tau^2} + \frac{dy_0}{d\tau} = 0, \quad y_1(0) = 0, \qquad \frac{dy_1}{d\tau}(0) = 0$$

$$O(\varepsilon^2) : \frac{d^2 y_2}{d\tau^2} + \frac{dy_1}{d\tau}, \qquad y_2(0) = 0, \qquad \frac{dy_2}{d\tau}(0) = 0 \tag{2.23}$$

$$O(\varepsilon^3) : \ldots$$

STEP D: Successively solve the sequence of equations obtained in (2.22):

$$y_0(\tau) = \tau - \tau^2/2, \qquad y_1(\tau) = -\tau^2/2 + \tau^3/6, \qquad y_2(\tau) = \tau^3/6 - \tau^4/24 \tag{2.24}$$

Substituting (2.24) into (2.21) gives (2.20)

Exercise 1: Show that $y_n(\tau) = (-1)^n(\tau^{n+1}/(n+1)! - \tau^{n+2}/(n+2)!)$.

Exercise 2: The maximum altitude reached by the projectile is

$$y_* := y(\tau_*),$$

where τ_* is the time such that $dy(\tau_*)/d\tau = 0$ (why?). From (2.18) we have

$$\frac{dy}{d\tau} = \frac{(1+\varepsilon)e^{-\varepsilon\tau}}{\varepsilon} - \frac{1}{\varepsilon}$$

Hence

$$\tau_* = \frac{In(1+\varepsilon)}{\varepsilon},$$

and

$$y_* = y(\tau_*) = \frac{1}{\varepsilon} - \frac{In(1+\varepsilon)}{\varepsilon^2} \tag{2.25}$$

For small ε, τ_* and y_* can be approximated by (recall that $In(1+z) = z - z^2/2 + z^3/3 + O(z^4)$),

$$\tau_* = 1 - \varepsilon/2 + \varepsilon^2/2 + O(\varepsilon^3) \qquad y_* = 1/2 - \varepsilon/3 + \varepsilon^2/4 + O(\varepsilon^3) \tag{2.26}$$

Re-obtain these expansions from (2.19) by perturbation theory.

Exercise 3: Check that (2.25) is consistent with the result we obtained for h using dimensional analysis.

Exercise 4: Recall that in the high friction limit, the appropriate equation for the projectile is

$$\frac{d^2z}{ds^2} = \frac{dz}{ds} - \delta, \qquad z(0) = 0, \qquad \frac{dz}{ds}(0) = 1 \tag{2.27}$$

where $\delta := 1/\varepsilon = mg/(kV)$. Here the altitude is measured in units of $l_2 = Vm/k$, and the time is units of $\tau_2 = m/k$. Solve this equation by regular perturbation technique where $0 < \delta \ll 1$.
What is the maximal altitude reached by the projectile? What is the time of the flight? What is the ratio between the ascent and descent time of the projectile?

Exercise 5: The dimensionless form of the equation for a perfect pendulum of length l is

$$\frac{d^2\theta}{d\tau^2} = -\sin(\theta), \qquad \theta(0) = \phi, \qquad \frac{d\theta}{d\tau}(0) = 0,$$

where the time is measured in units of $\sqrt{l/g}$. Solve this equation by regular perturbation where $0 < \phi \ll 1$. Is your approximation uniformly valid in time?

2.3.3 Eigenvalue problem

We will find the first-order perturbation of the eigenvalues of the differential equation

$$y'' + \lambda y + \varepsilon y^2 = 0$$

in $0 < x < \pi$, with boundary conditions $y(0) = y(pi) = 0$.
[**Exercise: Repeat this along with the final term as εy (easy) or εy^3 (hard).**]
First, we look at the case $\varepsilon = 0$:

$$y'' + \lambda y = 0$$

This has possible solutions:

$$
\begin{aligned}
\lambda < 0 \qquad & y = A\cosh[x\sqrt{-\lambda}] + B\sinh[x\sqrt{-\lambda}] \\
\lambda = 0 \qquad & y = Ax + B \\
\lambda > 0 \qquad & y = A\cos[x\sqrt{\lambda}] + B\sin[x\sqrt{\lambda}]
\end{aligned}
$$

The first two solutions cannot satisfy boundary conditions. The third must have $A = 0$ to satisfy the condition $y(0) = 0$, and the second condition leaves us with

$$B\sin[\pi\sqrt{\lambda}] = 0 \qquad \Rightarrow \qquad \lambda = m^2, m = 1, 2, \ldots$$

Now we return to the full problem, posing regular expansions in both y and λ:

$$y = \sin mx + \varepsilon y_1 + \ldots$$

$$\lambda = m^2 + \varepsilon\lambda_1 + \ldots$$

Substituting, we obtain for the differential equation:

$$-m^2 \sin mx + m^2 \sin mx = 0$$

$$\varepsilon y_1'' + \varepsilon m^2 y_1 + \varepsilon\lambda \sin mx + \varepsilon \sin^2 mx = 0$$

As we would expect, the order 1 equation is already satisfied, along with the boundary conditions.

Order ε

The ODE at order ε becomes

$$y_1'' + m^2 y_1 = -\lambda_1 \sin mx - \sin^2 mx = -\lambda_1 \sin mx + \frac{1}{2}\cos 2mx - \frac{1}{2}$$

. We expect a solution to the form

$$y_1 = A\sin mx + B\cos mx + Cx\cos mx + D\cos 2mx + E$$

and substituting this form back into the left-hand side gives us

$$-2Cm\sin mx - 3m^2 D\cos 2mx + Em^2 = -\lambda_1 \sin mx + \frac{1}{2}\cos 2mx - \frac{1}{2}$$

which fixes $C = \lambda_1/2m, D = -1/6m^2, E = -1/2m^2$. The solution is

$$y_1 = A\sin mx + B\cos mx + \frac{\lambda_1}{2m}x\cos mx - \frac{1}{2m^2} - \frac{1}{6m^2}\cos 2mx.$$

Now we apply the boundary conditions to determine the eigenvalue: $y(0) = 0$ gives

$$0 = B - \frac{1}{2m^2} - \frac{1}{6m^2} \qquad\qquad B = \frac{2}{3m^2}$$

and then the condition $y(\pi) = 0$ becomes:

$$0 = \frac{2}{3m^2}(-1)^m + \frac{\lambda_1}{2m}\pi(-1)^m - \frac{1}{2m^2} - \frac{1}{6m^2}$$

which simplifies to determine λ_1:

$$\lambda_1 = \frac{4}{3m\pi}[(-1)^m - 1] = \frac{-8}{3m\pi}\begin{cases} 0, & m \text{ even} \\ 1, & m \text{ odd} \end{cases}$$

Thus the eigenvalues become

$$\lambda = 1 - \frac{8\varepsilon}{3\pi}, 4, 9 - \frac{8\varepsilon}{9\pi}, 16, 25 - \frac{8\varepsilon}{15\pi}, \dots$$

2.3.4 Fluid Mechanics Problem

For a flow between two parallel paths, the upper is in uniform motion, and the lower is at rest (also known as plane Couette Flow) is added with uniform suction in the lower plate.

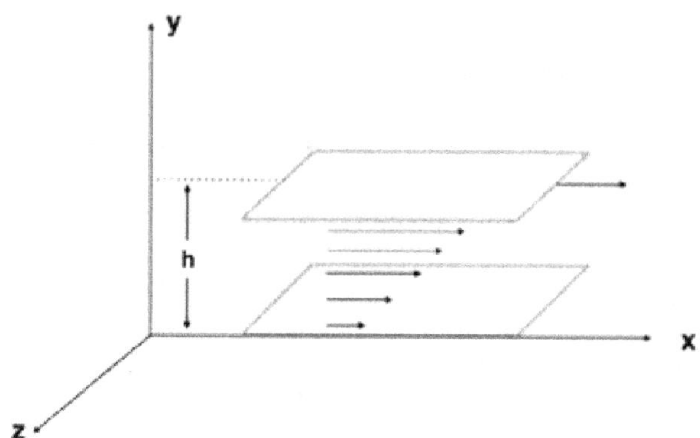

Figure 2.4: Plane Couette Flow. The upper plate is moving in the x-direction.

We assume two-dimensional laminar flow (i.e. $\dfrac{\partial \omega}{\partial z} = 0$) x-axis between two parallel plates (Fig. 2.4). As suction is uniform, v is independent of x i.e. $\dfrac{\partial v}{\partial x} = 0$. The equation of continuity for incompressible fluid is

$$\frac{\partial u}{\partial x} + \frac{\partial v}{\partial y} = 0 \tag{2.28}$$

Equation of motion is

$$u\frac{\partial u}{\partial x} + v\frac{\partial u}{\partial y} = -\frac{1}{\rho}\frac{\partial p}{\partial x} + \nu\frac{\partial^2 u}{\partial y^2}$$

$$v\frac{\partial v}{\partial y} = -\frac{1}{\rho}\frac{\partial p}{\partial y} + \nu\frac{\partial^2 v}{\partial y^2} \tag{2.29}$$

Let

$$\begin{cases} y = -y_0; v = -v_0; u = 0 \\ y = +y_0; v = +v_0; u = 0 \end{cases}$$

Let us introduce the following non-dimensional equalities:

$$\bar{u} = \frac{u}{U}, \bar{v} = \frac{v}{v_0}, \bar{x} = \frac{x}{y_0}, \eta = \frac{v}{y_0}, p = \frac{p}{v^2}, \sigma = \frac{v_0 y_0}{\nu}, R = \frac{v y_0}{\nu}$$

Hence in non-dimensional form, equation of continuity is

$$\frac{\partial \bar{u}}{\partial \bar{x}} + \frac{\sigma}{R}\frac{\partial \bar{v}}{\eta} = 0 \tag{2.30}$$

and equation of motion is

$$\bar{u}\frac{\partial \bar{u}}{\partial \eta} + \frac{\sigma \bar{v}}{R}\frac{\partial \bar{u}}{\partial \eta} = -\frac{\partial \bar{p}}{\partial \bar{x}} + \frac{1}{R}\frac{\partial^2 u}{\partial \eta^2}$$

$$\bar{v}\frac{\partial \bar{v}}{\partial \eta} = -\frac{R^2}{\sigma^2}\frac{\partial \bar{p}}{\partial \eta} + \frac{1}{\sigma}\frac{\partial^2 y}{\partial \eta^2} \tag{2.31}$$

Let

$$\begin{cases} \eta = 1; \bar{v} = -1; \bar{u} = 0 \\ \eta = 1; \bar{v} = +1; \bar{u} = 1 \end{cases}$$

We introduce perturbation terms (p', u', v') in the plane Couette Flow $(p_0, u_0, v = 0)$ as under

$$\bar{p}(\bar{x}, \eta) = p_0 + p'(\bar{x}, \eta)$$

$$\bar{u}(\bar{x}, \eta) = u_0 + u'(\bar{x}, \eta) \tag{2.32}$$

$$\bar{v}(\eta) = v'(\eta)$$

The plane Couette Flow satisfies the equations

$$\frac{\partial p_0}{\partial \eta} = v, \frac{\partial u_0}{\partial \bar{x}} = 0, \frac{\partial p_0}{\partial \bar{\eta}} = a_0 (constant)$$

$$and\, u_0 = \frac{1}{2}(\eta + 1) + A(\eta^2 - 1) \text{ where } A = \frac{Ra_0}{2} \tag{2.33}$$

Substituting eq. (2.32) and (2.33) in eq. (2.30) and eq. (2.31) we get

$$\frac{\partial u'}{\partial x} + \frac{\sigma}{R}\frac{\partial v'}{\partial \eta} = 0 \tag{2.34}$$

$$(u_0 + u')\frac{\partial u'}{\partial \bar{x}} + \frac{\sigma}{R}v'(\frac{1}{2} + 2A\eta + \frac{\partial u'}{\partial \eta}) = -\frac{\partial p'}{\partial \bar{x}} + \frac{1}{R}\frac{\partial^2 u'}{\partial \eta^2} - a_0 + \frac{2A}{R}$$

$$v'\frac{\partial v'}{\partial \eta} = -\frac{R^2}{\sigma^2}\frac{\partial p'}{\partial \eta} + \frac{1}{\sigma}\frac{\partial^2 v'}{\partial \eta^2} \tag{2.35}$$

The boundary conditions are

$$\eta = -1, v' = -1, u' = 0$$

$$\eta = +1, v' = +1, u' = 1$$

Now let $v' = -f(\eta)$ then eq. (2.34) gives

$$u' = \frac{\sigma}{R}\bar{x}f'(\eta) + F(\eta)$$

We assume $f'(\eta) = \dfrac{\partial f}{\partial \eta}$, $p' = \dfrac{\partial f}{\partial \eta}$

where $t(\eta)$ and $F(\eta)$ are the unknown functions to be determined. Substituting values of u' and v' in eq. (2.34) and in eq. (2.35) we get

$$\frac{\partial p'}{\partial \eta} = -\frac{\sigma}{R^2}(\sigma f.f' + f'')$$

and

$$\frac{\partial p'}{\partial \bar{x}} = \frac{1}{R}\{F'' + 2A - \sigma(\frac{1}{2}f' + \frac{1}{2}f'\eta + A\eta^2 f' - Af' - \frac{1}{2}f - \\ 2A\eta f + f'F - fF')\} + \frac{\sigma}{R^2}\bar{x}\{f''' - \sigma(f'^2 - ff'')\} - a_0 \tag{2.36}$$

Now at the mouth of the channel, the pressure gradient along the axis of the channel is assumed to be the same as the pressure gradient of the flow without porosity.

Thus, $\dfrac{\partial p_0}{\partial \bar{x}} = \dfrac{\partial \bar{p}}{\partial \bar{x}} = (a_0)$

Say for, at $\bar{x} = 0$, $\dfrac{\partial p'}{\partial \bar{x}} = 0$.

Hence from above equation we have,

$$F'' - \sigma(1/2f' + 1/2\eta f' + A\eta^2 f' - Af' - 1/2f - 2A\eta f + f'F - fF')$$
$$= (a_0 R - 2A) = 0 \tag{2.37}$$

Differentiating eq. (2.36) with respect to $\bar{x}$, we get,

$$\frac{\partial^2 p'}{\partial \bar{x}\partial \eta} = 0 \tag{2.38}$$

Again differentiating eq. (2.36) with respect to η, and simplifying, by using the results (2.37) and (2.38), we have,

$$\frac{d}{d\eta}[f''' - (f'^2 - ff'')] = 0 \tag{2.39}$$

which is true for all $\bar{x}$.
After integrating (2.39) we get

$$f''' - \sigma(f'^2 - ff'') = c \tag{2.40}$$

where c is the constant of integration. The boundary condition becomes,

$$\eta = -1, f(-1) = 1, f'(-1) = 0, F(-1) = 0$$

$$\eta = +1, f(1) = -1, f'(1) = 0, F(1) = 0$$

In order to solve the equation (2.40), let us consider the power series for f and c near $\sigma = 0$ as $(0.01 \leq \sigma \leq 0.1)$

$$f = f_0 + \sigma f_1 + \sigma^2 f_2 + \cdots + \sigma^n f_{n'} \tag{2.41}$$

and

$$c = c_0 + \sigma c_1 + \sigma^2 c_2 + \cdots + \sigma^n c_{n'} \tag{2.42}$$

where f_n and c_n are independent of σ. By substituting equation (2.41) and (2.42) into (2.40) and equating the like powers of σ we get,

$$f_0''' = c_0 \tag{2.43}$$

$$f_1''' - f_0'^2 + f_0 f_0'' = c_1 \tag{2.44}$$

$$f_2''' - 2f_0' f_1' + f_1 f_0'' + f_0 f_1'' = c_2 \tag{2.45}$$

In order to solve (2.43), (2.44) and (2.28) the boundary condition for f_n are,

$$f_0(-1) = 1, f_n(-1) = 0 \text{ for } n \geq 1,$$

$$f_n'(-1) = 0 \text{ for } n \geq 0,$$

$$f_0(1) = -1, f_n(1) = 0 \text{ for } n \geq 1,$$

$$f_n'(1) = 0 \text{ for } n \geq 0.$$

Thus second-order perturbation solution of (2.40) is

$$\begin{aligned}
f(\eta) = &\frac{1}{2}(\eta^3 - 3\eta) + \frac{\sigma}{280}(\eta^7 - 3\eta^3 + 2\eta) \\
&- \frac{\sigma^2}{123600}(14\eta^{11} - 385\eta^9 + 19\eta^7 + 836\eta^3 - 703\eta)
\end{aligned} \tag{2.46}$$

and

$$c = 3 - \frac{81}{35}\sigma + \frac{234}{13475}\sigma^2 \tag{2.47}$$

Similarly, to solve (2.37) we consider the power series for F near $\sigma = 0$ as

$$F = F_0 + \sigma F_1 + \sigma F_2 + \cdots + \sigma^n F_n \tag{2.48}$$

Substituting in (2.37), (2.46) and (2.48), we get after equating the like power of σ,

$$F_0'' = 0 \tag{2.49}$$

$$F_1'' - (1/2 + 1/2\eta + A\eta^2 - A)F f_0' + (\frac{1}{2} + 2A\eta)f_0 - f_0' F_0 + f_0 F_0' = 0 \tag{2.50}$$

$$F_2'' - (1/2 + 1/2\eta + A\eta^2 - A)f_1' + (\frac{1}{2} + 2A\eta)f_1 - f_0'F_1 - f_1'F + f_0F_1' + f_1F_0' = 0$$

$$(2.51)$$

and boundary conditions are

$$\eta = -1, F_n(-1) = 0 \; for \; n \geq 0$$

and

$$\eta = 1, F_n(1) = 0 \; for \; n \geq 0.$$

Thus the second-order perturbation solution of equation (2.37) is

$$F = \frac{\sigma}{240}[44\eta^6 + 45\eta^2 - 46 + 3(2\eta^5 + 5\eta^4 - 30\eta^2 - 2\eta + 25)]$$

$$- \frac{\sigma^2}{2822400}[8A(28\eta^{10} - 630\eta^8 - 4284\eta^6 + 16905\eta^4 - 303030\eta^2$$

$$+ 2991011) + 35(16\eta^9 + 27\eta^8 - 288\eta^7 - 252\eta^6 + 144\eta^5 + 3888\eta^4$$

$$+ 18756\eta^2 + 128\eta - 2241)]$$

The velocity components in $\bar{x}$ and η directions are

$$u = \frac{1}{2}(\eta + 1) + A(\eta^2 - 1) + \frac{\sigma\bar{x}}{R}[\frac{3}{2}(\eta^2 - 1) + \frac{\sigma}{280}(7\eta^6 - 9\eta^2 + 2)$$

$$- \frac{\sigma^2}{1293600}(154\eta^{10} - 3465\eta^8 + 1386\eta^6 + 2628\eta^2 - 703)]$$

$$+ \frac{\sigma}{240}[4A(\eta^6 + 45\eta^2 - 46) + 3(2\eta^5 + 5\eta^4 - 30\eta^2 - 2\eta + 25)]$$

$$- \frac{\sigma^2}{2822400}[8A(28\eta^{10} - 630\eta^8 - 4284\eta^6 + 169905\eta^4 - 303030\eta^2$$

$$+ 291011) + 35(16\eta^9 + 27\eta^8 - 288\eta^7 - 252\eta^6 + 144\eta^5 + 3888\eta^4$$

$$+ 18756\eta^2 + 128\eta - 22419)]$$

and

$$\bar{v} = -\frac{1}{2}(\eta^3 - 3\eta) - \frac{\sigma}{280}(\eta^7 - 3\eta^3 + 2\eta)$$

$$+ \frac{\sigma^2}{1293600}(14\eta^{11} - 385\eta^9 + 198\eta^7 + 876\eta^3 - 703\eta)$$

The pressure distribution is given by the expression,

$$\bar{p}(0,0) - \bar{p}(\bar{x},\eta) = \frac{\sigma}{R^2}(f' + \frac{3}{2}) - \frac{\sigma^2}{R^2}(\frac{1}{140} - \frac{f^2}{2}) - \frac{703}{1293600R^2} - \frac{\sigma\bar{x}^2c}{2R^2}$$

Chapter 3

Singular Perturbation Theory

It concerns the study of problems featuring a parameter for which the solution of the problem at a limiting value of the parameter is different in character from the limit of the solution of the general problem. For regular perturbation problems, the solution of the general problem converges to the solution of the limit problem as the parameter approaches the limit value.

Example - 3.1: Consider

$$\epsilon x^2 + x + 1 = 0 \tag{3.1}$$

Since equation (3.1) is a quadratic equation, it has two roots. For $\epsilon \to 0$ Equation (3.1) reduce to

$$x + 1 = 0 \tag{3.2}$$

which is of the first order. Thus x is discontinuous at $\epsilon = 0$. Such perturbation is called *singular perturbation problem..*

$$x = x_0 + \epsilon x_1 + \epsilon^2 x_2 + \ldots \tag{3.3}$$

Putting this value in equation (3.1)

$$\epsilon(x_0 + \epsilon x_1 + \ldots) + x_0 + \epsilon x_1 + \cdots + 1 = 0$$
$$\Rightarrow \epsilon(x_0^2 + 2\epsilon x_0 x_1 + \ldots) + x_0 + \epsilon x_1 + \cdots + 1 = 0$$
$$\Rightarrow \epsilon x_0^2 + 2\epsilon^2 x_0 x_1 + \cdots + x_0 + \epsilon x_1 + \cdots + 1 = 0$$
$$\Rightarrow \epsilon(x_0^2 + x_1) + x_0 + 1 = 0$$

Equating the co-efficient of the like power of ϵ gives

$$x_0 + 1 = 0$$
$$x_1 + x_0^2 = 0$$

when $x_0 = -1, x_1 = -1$ so one of the root is

$$x = -1 - \epsilon + \ldots \tag{3.4}$$

Thus as expected the above procedure yielded only one root. We investigate the exact solution i.e.,

$$x = \frac{1}{2\epsilon}(-1 \pm \sqrt{1 - 4\epsilon}) \tag{3.5}$$

Using the binomial theorem, we have

$$\sqrt{1 - 4\epsilon} = 1 - 2\epsilon + \frac{(\frac{1}{2})(\frac{-1}{2})}{2!} \times (-4\epsilon)^2 + \ldots$$
$$= 1 - 2\epsilon - 2\epsilon^2 + \ldots \tag{3.6}$$

Substituting (3.6) in (3.5)

$$x = \frac{-1 + 1 - 2\epsilon - 2\epsilon^2 + \ldots}{2\epsilon} = -1 - \epsilon + \ldots \tag{3.7}$$
$$x = \frac{-1 - 1 + 2\epsilon + 2\epsilon^2 + \ldots}{2\epsilon} = \frac{-1}{\epsilon} + 1 + \epsilon + \ldots \tag{3.8}$$

Therefore, both of the roots go in the power of ϵ but one starts with ϵ^{-1}. Hence it is not surprising that the assumed expansion in (3.3) is failed to produce the root (3.8). Consequently one can not determine the second root of the equation is not to be reduced, the order tends to ∞ as $\epsilon \to 0$ and hence, assume that the leading term has the form

$$x = \frac{y}{\epsilon^v} \tag{3.9}$$

where v must be greater than zero and need to be determined in the course of analysis. Substituting (3.9) in (3.1)

$$\epsilon^{1-2v} y^2 + \epsilon^v y + 1 + \cdots = 0$$

Since $v > 0$, the second term is much bigger than 1. Hence the dominant part of (3.9) is

$$\epsilon^{1-2v} y^2 + \epsilon^v y = 0 \tag{3.10}$$

which demands that power of ϵ be the same.

$$1 - 2v = -v \Rightarrow v = 1$$

For $v = 1 \Rightarrow y = 0 \, or - 1$.

The first value $y = 0$, corresponds to the first root $x = -1 - \epsilon$. For $y = -1$, it corresponds to second root. Thus it follows from (3.9)

$$x = \frac{-1}{\epsilon} + \dots$$

To determine more terms in the expansion of second root, we try

$$x = \frac{-1}{\epsilon} + x_0 + \dots \tag{3.11}$$

Substituting it in equation (3.1)

$$\Rightarrow \epsilon \left(\frac{-1}{\epsilon} + x_0 + \dots \right)^2 - \frac{-1}{\epsilon} + x_0 + \dots + 1 = 0$$

$$\Rightarrow \epsilon \left(\frac{-1}{\epsilon} + \frac{2x_0}{\epsilon} + x_0^2 + \dots \right) - \frac{-1}{\epsilon} + x_0 + 1 + \dots = 0$$

$$\Rightarrow -2x_0 + x_0 + 1 + O(\epsilon) = 0$$

$\Rightarrow x_0 = 1$ and equation (3.11) becomes

$$x = -\frac{1}{\epsilon} + 1 + \dots$$

Alternatively, once v has been determined. We view (3.9) as a transformation from x to y. Then putting $x = \frac{y}{\epsilon}$ in (3.1) yields,

$$y^2 + y + \epsilon = 0 \tag{3.12}$$

Which can be solved to determine both the roots because ϵ does not multiply the highest order.

Using binomial theorem, we have

$$(1 + 8\epsilon + 4\epsilon^2)^{\frac{1}{2}} = 1 + \frac{1}{2}(8\epsilon + 4\epsilon^2) + \frac{(\frac{1}{2})(\frac{-1}{2})}{2!}(8\epsilon + 4\epsilon^2)^2 + \dots$$

$$= 1 + 4\epsilon + 2\epsilon^2 - \frac{1}{8}(64\epsilon^2 + \dots)$$

$$= 1 + 4\epsilon + 2\epsilon^2 - 8\epsilon^2 + \dots$$

$$= 1 + 4\epsilon - 6\epsilon^2 + \dots$$

Further, we have

$$x = \frac{1}{2}(3 + 2\epsilon + 1 + 4\epsilon - 6\epsilon^2 + \dots)$$
$$= 2 + 3\epsilon - 3\epsilon^2 + \dots$$
$$x = \frac{1}{2}(3 + 2\epsilon - 1 - 4\epsilon + 6\epsilon^2 + \dots)$$
$$= 1 - \epsilon + 3\epsilon^2 + \dots$$

Which are same as equation (3.11) and (3.12).

When SSA i.e., steady state approximation is applied in chemical kinetics, the argument is that the fraction of the intermediates is very reactive and hence they are eliminated as fast as they are made. Setting the corresponding rate of change to zero, by this we mean that they are not identically zero but this rate of change is much slower than the other rates of reaction.

Surprisingly, SSA is often accurate yet lacks rigor in the classical arguments. Hence, it does not lead to any satisfying mathematical generalization which can be helpful for the improvement of approximation. The following two mathematical theories, built on SSA can obtain the required generalization:

1. Singular Perturbation theory: General solutions of equations in perturbation theory consists of power series in small parameter. In SSA, we come across problems where the small parameter is multiplied by a derivative.

2. Slow manifold theory: Slow invariant manifolds of the differential equations can be thought of as approximations of the curves or surfaces that result from the SSA.

The given two approaches are related.

3.1   Scaling and balancing

One issue with the traditional SSA arguments is that they leave us in a difficult position when it comes to establishing quantitative standards for their validity. Take the Michaelis-Menten mechanism, for instance:

$$E + S \underset{k_{-1}}{\overset{k_1}{\rightleftharpoons}} C \overset{k_{-2}}{\rightarrow} E + P$$

E is the enzyme, S is the substrate, P is the product, and C is the intermediate (enzyme substrate/product) combination in this mechanism.

The standard defense goes something like this:[1] Enzymes are catalysts, and they work quite well. As a result, we anticipate that the conversion of C into enzyme and product will happen quickly. Fast in relation to what? Runs the typical response in comparison to the creation of the enzyme-substrate complex. This reasoning is valid up to a point, but what does it mean to argue that one chemical reaction is quicker than another? Which figures are we contrasting? Because the rates of formation and degradation are roughly equal in the steady state, it cannot be rates. Due to the fact that k_1 has different units than k_{-1} and k_{-2}, it cannot be rate constants.

Until we understand that we're going to run into problems *as long as we try to work with values that have units, there is no easy way out of this problem.* In order to make accurate claims about what it means for one process to be faster than another, we might be able to find things to compare if we could do away with the units.

Write down the rate equations for this mechanism first:

$$\frac{dE}{dt} = -k_1 ES + k_{-1}C + k_{-2}C, \tag{3.13a}$$

$$\frac{dS}{dt} = -k_1 ES + k_{-1}C, \tag{3.13b}$$

$$\frac{dC}{dt} = k_1 ES - k_{-1}C - k_{-2}C, \tag{3.13c}$$

$$\frac{dP}{dt} = k_{-2}C. \tag{3.13d}$$

The two equations result from mass conservation.

$$E_0 = E + C$$
$$S_0 = S + C + P$$

Because $\dfrac{dE}{dt} + \dfrac{dC}{dt} = \dfrac{dS}{dt} + \dfrac{dC}{dt} + \dfrac{dP}{dt} = 0$, it should be noted that these conservation rules are the results of the differential equations. Therefore, there is no difference between the solutions when using these algebraic equations. Only the set of differential equations 1 is reduced to the equivalent planar system.

---

[1]The classical argument can be put a little better than this, but the underlying problem remains.

$$\frac{dS}{dt} = -k_1 S(E_0 - C) + k_{-1}C, \tag{3.14a}$$

$$\frac{dC}{dt} = k_1 S(E_0 - C) - (k_{-1} + k_{-2})C, \tag{3.14b}$$

The SSA is based on the observation that the concentrations of highly reactive intermediates, such as C, eventually approach a rough equilibrium between production and destruction, resulting in a minimal net rate of change for C. That balance must be clear, is what we desire. To do this, we must first select new measurement scales that will allow us to measure each of the variables (S, C, and t) on a scale of unit magnitude. In other words, by using the new variables s, c, and τ

$$s = S/\tilde{S}, \tag{3.15a}$$

$$c = C/\tilde{C}, \tag{3.15b}$$

$$and \quad \tau = t/\tilde{t}, \tag{3.15c}$$

where the tilde designates a measuring scale whose units are the same as those of the corresponding variable and such that, to the extent that this is feasible, the new variables are of unit magnitude. S is simple to select: The majority of experiments will begin with $S(0) = S_0$, thus must be a favorable choice.

$$\tilde{S} = S_0 \tag{3.16}$$

should be an appropriate choice.

Though challenging, choosing $\tilde{C}$ is still not that difficult. As is customary in experiments, if we begin at the initial point $(S, C) = (S_0, 0)$, C will first climb to some maximum before falling towards zero (the equilibrium concentration). From calculus we know that the function $C(t)$ will reach its maximum when $dC/dt = 0$, or in other words when,

$$C = \frac{k_1 E_0 S}{k_1 S + k_{-1} + k_{-2}} \tag{3.17}$$

It is important to note that this is the standard SSA. If we knew the value of S at the point where C reaches its maximum, we could use that number as $\tilde{C}$. However, there is no simple method for calculating this value of S. Having stated that we do not need to know C's upper bound precisely. Really, all we need for the scale factor is an approximation of the order of magnitude. Early in the reaction, before much S has been consumed, is when we anticipate the maximum in C to be obtained. $S = S_0$ is consequently substituted into equation 3.17 to obtain,

$$\tilde{C} = \frac{k_1 E_0 S_0}{k_1 S_0 + k_{-1} + k_{-2}} = \frac{E_0 S_0}{S_0 + K_S} \tag{3.18}$$

where the Michaelis constant, which is well-known from classical enzyme kinetics, is $K_S = (k_{-1} + k_{-2})/k_1$.

Since it's not entirely clear what we should use for $\tilde{t}$, for now, we'll just use the scaling factors 4 and 6 to enter equations 3.15 into the rate equations 3.14 and see what we get

$$S = s S_0$$

$$C = c\tilde{C} = c\frac{E_0 S_0}{S_0 + K_S}$$

$$t = \tau \tilde{t}$$

$$\therefore \frac{dS}{dt} = \frac{d(s S_0)}{d(\tau \tilde{t})}$$

$$= -k_1 s S_0 \left(E_0 - c\frac{E_0 S_0}{S_0 + K_S} \right) + k_{-1} c\frac{E_0 S_0}{S_0 + K_S}$$

$$\therefore \frac{dS}{d\tau} = \tilde{t} \left\{ -k_1 E_0 s \left(1 - c\frac{S_0}{S_0 + K_S} \right) + k_{-1} c\frac{E_0}{S_0 + K_S} \right\}$$

Similarly for C:

$$\frac{dc}{d\tau} = \tilde{t} k_1 (S_0 + K_S) \left\{ s \left(1 - c\frac{S_0}{S_0 + K_S} \right) - c\frac{K_S}{S_0 + K_S} \right\}$$

Let

$$\alpha = \frac{S_0}{S_0 + K_S}$$

The value of this parameter ranges from 0 to 1. We get two rate equations

$$\frac{ds}{d\tau} = \tilde{t} \left\{ -k_1 E_0 s(1 - \alpha c) + k_{-1} c\frac{E_0}{S_0 + K_S} \right\}$$

$$\frac{dc}{d\tau} = \tilde{t} k_1 (S_0 + K_S) \left\{ s(1 - \alpha c) - c(1 - \alpha) \right\}$$

The second of these equations' terms in the brace is now clearly balanced: Each is composed of a product of quantities that have unit magnitude. Only one option is left to select. $\tilde{t}$.

The back reaction $C \to E + S$ isn't extremely important for many enzymes. Below these conditions, the first term, $-k_1 E_0 s(1 - \alpha c)\tilde{t}$ ought to dominate ds/dt. This implies that we ought to take

$$\tilde{t}^{-1} = k_1 E_0,$$

to strike a balance between the term and its derivative. To put it another way, we anticipate that $(k_1 E_0)^{-1}$ will indicate a slow pace time scale that regulates how quickly the substrate is consumed. Obviously, we should choose a different value for $\tilde{t}$ if the aforementioned hypothesis is incorrect for a particular enzyme. In any case, our decision results in

$$\frac{ds}{d\tau} = -s(1 - \alpha c) + c\frac{k_{-1}}{k_{-1}k_{-2}}(1 - \alpha)$$

$$\frac{dc}{d\tau} = \frac{S_0 + K_S}{E_0}\{s(1 - \alpha c) - c(1 - \alpha)\}$$

Define

$$\beta = \frac{k_{-1}}{k_{-1} + k_{-2}}$$

and

$$\mu = \frac{E_0}{S_0 + K_S}$$

Then we have

$$\frac{ds}{d\tau} = -s(1 - \alpha c) + \beta c(1 - \alpha), \qquad (3.19a)$$

$$and \qquad \mu\frac{dc}{d\tau} = s(1 - \alpha c) - c(1 - \alpha). \qquad (3.19b)$$

The SSA is said to be valid when the enzyme concentration is low in traditional enzymology. This relates to a negligible value of μ. This statement is rationale and a specific requirement for its validity are both provided by the aforementioned scaling: The differential equation 3.19b is "close" to the algebraic equation if μ is very small.

$$0 \approx s(1 - \alpha c) - c(1 - \alpha), \qquad (3.20)$$

i.e. the SSA. A small value of μ implies that

$$E_0 \ll S_0 + K_S,$$

i.e, SSA will be valid in this circumstance. It should be noted that this is a sufficient condition, not a necessity. We might discover more factors that contribute to the validity of the SSA if we scaled time differently.

Take note that there is a humorous element to what we are doing. We began with two differential equations and concluded that one of the differential equations degenerates into an algebraic equation. A system of two differential equations differs significantly from a single differential equation with an algebraic constraint in that the former only needs a value for s, while the latter computes the value of c using equation 3.20. By taking into account the limit $\mu \to 0$, which can never be achieved in practice, we have altered the nature of the problem. These problems are known as **singular perturbation problems**. Singular perturbation problems can be exceedingly challenging since the nature of the problems we started with and the ones we obtain in the limit where our small parameter reaches zero are different. However, they appear frequently in the sciences, so it is worthwhile to give them some additional thought.

The usage of the limit $\mu = 0$ to approximate the slow manifold (which is what we obtain from equation 3.20 if we solve for c) is supported by a theorem regardless of the unusual nature of singular perturbation problems: Think about a set of differential equations of the following type.

$$\frac{d\mathbf{x}}{dt} = \mathbf{f}(\mathbf{x},\, \mathbf{z}, t), \tag{3.21a}$$

$$\mu \frac{d\mathbf{z}}{dt} = \mathbf{g}(\mathbf{x},\, \mathbf{z}, t). \tag{3.21b}$$

In the singular limit $\mu = 0$, we obtain the system

$$\frac{d\mathbf{x}}{dt} = \mathbf{f}(\mathbf{x},\, \mathbf{z}, t),$$

$$\mathbf{z} = \phi(\mathbf{x}, t),$$

where the answer to $g(x, z, t) = 0$ is the second of these equations. This system is referred to as **degenerate system**. The **adjoined system** is what we define last, and it is merely, define the **adjoined system**, which is just

$$\frac{dz}{dt} = \mathbf{g}(\mathbf{x},\, \mathbf{z}, t)$$

now we treat x and t as constants. [2]

Theorem 1 (Tikhonov's theorem) *When $\mu \to 0$, the solution of the system 3.21 tends to the solution of the degenerate system if $z = f(x,t)$ is a stable root of the adjoined system.*

(Although it should be done more often, the stability of the adjoined system is rarely evaluated.) As a result of Tikhonov's theorem, there is

---

[2]If you're wondering where the μ went in the adjoined system, the time variable can be rescaled in such a way as to make this coefficient disappear. Since stability doesn't depend on how we scale time, there's no point carrying this parameter around.

always a slow manifold whose approximate formula is given by the SSA for sufficiently small μ. The good review study by Klonowski provides more information.

Imagine starting an experiment for a system with a low value of μ at the point $(s, c) = (1, 0)$. We shall initially have $\dot{c} \equiv dc/d\tau = 1/\mu$ since $c = 0$. This is a large, positive quantity (since μ is small). As a result, the intermediate complex concentration will initially grow quickly, with a typical rise time of. Nevertheless, as c increases, $\dot{c}$ will contract. Since is small, equation (3.19b) right-hand side should also be small as a result, making equation (3.20) valid. In other words, we may divide the reaction's time course into two parts:

1. The chemical kinetics terms for the abrupt increase in the intermediate concentration include induction period and **transient**. This is referred to as the **inner solution** in the theory of singular perturbations. [3]

2. A more gradual approach towards equilibrium, during which the SSA is at least approximately valid. The external solution is this.

3.2   The outer solution

In this section, we find the outer solution to the lowest order in μ in this part. This use simple methods that you are supposed to have learned in your undergraduate classes. It is possible to obtain approximations of higher order. For more information, see Heineken, Tsuchiya, and Aris [2]

Solving equation (3.20) for c, we get

$$c = \frac{s}{\alpha s + 1 - \alpha} \tag{3.22}$$

In the differential equation (3.19a) with this equation as a substitute, we obtain

$$\begin{aligned}
\frac{ds}{d\tau} &= -s\left(1 - \alpha \frac{s}{\alpha s + 1 - \alpha}\right) + \beta(1 - \alpha)\frac{s}{\alpha s + 1 - \alpha} \\
&= -\frac{s(1 - \alpha)(1 - \beta)}{\alpha s + 1 - \alpha}
\end{aligned}$$

---

[3]The terms "inner solution" and "outer solution" are such for historical reasons: Singular perturbation problems were first considered in fluid dynamics, where the inner solution referred to the behavior near a wall, while the outer solution referred to the behavior away from the wall.

By separating the variables, this problem can be quickly solved:

$$\int_{s_0}^{s} ds \frac{\alpha s + 1 - \alpha}{s} = -(1-\alpha)(1-\beta) \int_{\tau_0}^{\tau} d\tau$$

$$\therefore \quad -(1-a)(1-b)(t-t_0) = \int_{s_0}^{s} ds \left(\alpha + \frac{1-\alpha}{s} \right)$$

$$= a(s-s_0) + (1-a)ln(s/s_0)$$

After the system has reached the manifold, its state is represented by the constants s_0 and t_0. Therefore, these are arbitrary values that are not connected (at least not directly) to the system's actual initial conditions. These constants can consequently be combined into a single, arbitrary constant A:

$$-(1-\alpha)(1-\beta)\tau + A = \alpha s + (1-\alpha)lns. \tag{3.23}$$

We will determine the constant A later.

3.3 The inner solution

Finding the inner solution is a little more difficult. The incorrect scaling of equation 3.19 is the issue. The rise time for the intermediate complex is μ, which is quite small given our time scaling. We must rescale the equations so that the rise time is not small in order to analyse the inner solution. The simplest method is as follows

$$\tau = \mu\theta$$

The new time variable θ is stretched by a factor of $1/\mu$ relative to τ. Our rate equations become

$$\frac{ds}{d\theta} = \mu[-s(1-\alpha c) + \beta c(1-\alpha)] \tag{3.24a}$$

$$and \quad \frac{dc}{d\theta} = s(1-\alpha c) - c(1-\alpha) \tag{3.24b}$$

When we set $\mu = 0$, or to the same degree of approximation as our outer solution, we obtain

$$\frac{ds}{d\theta} = 0$$

keeping equation (3.24b) intact. As a result, we can assume that s is approximately constant throughout the induction period, at least roughly speaking. We define s_{in} as the inner solution, setting it apart from the outer solution. $s_{in} = 1$ to lowest order in μ because $s = S/S_0$ and S_0 is the initial concentration of S.

With $s = s_{in} = 1$, equation (3.24b) reduces to

$$\frac{dc}{d\theta} = 1 - c$$

This simple linear differential equation can be integrated starting with the initial value of $c = 0$:

$$\int_0^c \frac{dc}{1-c} = \int_0^\theta d\theta$$
$$\therefore \theta = -In(1-c), \tag{3.25}$$
$$or \quad c = 1 - e^{-\theta}$$

3.4 Matching the inner and outer solutions

The inner and outer solutions must coincide since they only depict two segments of the same trajectory. We shall be able to determine the parameter A given those conditions. The inner solution is valid when τ isn't too large, but the outer solution is believed to be valid for values of τ that aren't too small. We take into account both small values of τ and big values of θ in order to match the two answers. While large values of θ should correspond to the extremely late stages of the induction period, small values of τ should represent the outer solution's early moments following the transient's decay. If everything goes according to plan, we should be able to compare the two solutions under this intermediary regime. If we let $\tau|0$ and $s \to s_{in}$ in the outer solution 3.23, we get

$$A = \alpha s_{in} + (1 + \alpha)Ins_{in} = \alpha$$

Thus, s satisfies

$$-(1-\alpha)(1-\beta)\tau + \alpha = \alpha s + (1-\alpha)Ins \tag{3.26}$$

in the outer solution region.

No further information is obtained in this situation by matching the values of c: 3.25 yields $c = 1$ if we enter $s = s_{in} = 1$. By using the limit $\tau \to \infty$ in equation 3.25, we get $c \to 1$. As a result, the solution's c components coincide automatically.

The main approach for performing singular perturbation work is always, at least in outline, as follows:

1. Construct a suitable scale for the equations. Determine the small parameter(s).

2. Find the outer solution.

3. Find the inner solution.

4. Align the two solutions over an intermediate window of time.

The only snags are that we might want to develop the solutions to a higher order than we have done here and that the matching technique can be challenging to properly implement at times.

3.5 Geometric singular perturbation theory and the outer solution

The system's temporal evolution was the focus of our work in the earlier sections. For instance, 3.26 connects the time τ to s in the outer solution. The slow manifold, a structure that is best understood in phase space, is tied to these ideas, as we have already intimated. By formulating a singular perturbation expansion of the slow manifold, we now make this connection obvious. **Geometric singular perturbation theory**, which was developed by Fenichel, offers the theoretical foundation for carrying out this task.

Note that equation 3.22 implies that the manifold can be expressed as $c = c(s)$. Therefore, the manifold equation for our system would be

$$\frac{dc}{d\tau} = \frac{dc}{ds}\frac{ds}{d\tau}$$

Equations 3.19 have given us dc/dt and ds/dt . We know that the manifold simplifies the form 3.22 for small μ. This knowledge can be applied, for example, by writing the manifold as a power series in μ as shown in the following equation.

$$c(s) = \sum_{i=0}^{\infty} \gamma_i(s)\mu^i. \tag{3.27}$$

We should discover (and shall verify) that equation 3.22 provides the value of γ_0. We insert the power series 3.27 into the manifold equation in order to determine the unidentified functions $\gamma_i(s)$ (referred to as simply γ_i from this point forward). The procedure will be carried out in the following steps:

$$\frac{dc}{ds} = \sum_{i=0}^{\infty} \frac{\gamma_i}{ds} \mu^i.$$

$$\frac{dc}{ds} = \frac{1}{\mu} \left\{ s \left(1 - \alpha \sum_{i=0}^{\infty} \gamma_i \mu^i \right) - (1 - \alpha) \sum_{i=0}^{\infty} \gamma_i(s) \mu^{i-1} \right\}$$

$$= \frac{s}{\mu} - \alpha s \sum_{i=0}^{\infty} \gamma_i(s) \mu^{i-1} - (1 - \alpha) \sum_{i=0}^{\infty} \gamma_i(s) \mu^{i-1}$$

$$= \frac{s}{\mu} - (\alpha s + 1 - \alpha) \sum_{i=0}^{\infty} \gamma_i(s) \mu^{i-1}$$

$$\frac{ds}{d\tau} = -s \left(1 - \alpha \sum_{i=0}^{\infty} \gamma_i(s) \mu^i \right) + \beta(1 - \alpha) \sum_{i=0}^{\infty} \gamma_i(s) \mu^i$$

$$= -s + [\alpha s + \beta(1 - \aleph)] \sum_{i=0}^{\infty} \gamma_i(s) \mu^i$$

$$\therefore \quad \frac{s}{\mu} - (\alpha s + 1 - \alpha) \sum_{i=0}^{\infty} \gamma_i(s) \mu^{i-1}$$

$$= \sum_{i=0}^{\infty} \frac{d\gamma_i}{ds} \mu^i \left\{ -s + [\alpha s + \beta(1 - \alpha)] \sum_{i=0}^{\infty} \gamma_i \mu^i \right\} \tag{3.28}$$

We now aim at collecting terms with powers of μ. These terms might be read straight from the equation. The leading terms, for instance, are the μ^{-1} terms:

$$\frac{1}{\mu} \{ S - (\alpha S + 1 - \alpha) \gamma_0 \} = 0$$

Since this equation must be true for any $\mu \neq 0$, the term in braces must be zero, i.e.

$$\gamma_0 = \frac{s}{\alpha s + 1 - \alpha}, \tag{3.29}$$

which is precisely what we recommended that the term be. As long as we're only interested in small powers, this is not too difficult to accomplish. However, it is convenient to rewrite equation 3.28 so that each side appears as a power series in μ, rather than as a complex expression involving, among other things, a product of series, if we want to develop a general equation that would allow us to obtain the series expansion of the manifold to arbitrary order. Additionally, it is more convenient to write these series as sums involving μ^i rather than μ^{i-1}, as shown on the left, or other similar expressions. You may have studied the rules for rewriting sums in one of your maths classes. If not, the examples that follow should help you understand the reasoning behind these rules, which isn't too difficult once you've completed a few examples.

It's not too difficult to rewrite equation 3.28's first sum in the appropriate form

$$\sum_{i=0}^{\infty} \gamma_i \mu^{i-1} = \sum_{i=-1}^{\infty} \gamma_{i+1} \mu^{i}.$$

Try outlining the first few terms of each side of this identity if you are having trouble seeing it right away.

The right-hand side of equation 3.28's product of sums is the trickiest part. Here is how it changed:

$$\left(\sum_{j=0}^{\infty} \frac{d\gamma_j}{ds} \mu^j \right) \left(\sum_{k=0}^{\infty} \gamma_k \mu^k \right) = \sum_{i=0}^{\infty} \mu^i \sum_{j=0}^{\infty} \frac{d\gamma_j}{ds} \gamma_{i-j}.$$

The left-side summation indicators have been renamed as a convenience for clarity. The most important finding is that, when $j + k = i$, multiplying the two sums results in terms in μ_i.

Now that we're ready, let's return to equation 3.28:

$$\frac{s}{\mu} - (\alpha s + 1 - \alpha) \sum_{i=-1}^{\infty} \gamma_{i+1} \mu^i = -s \sum_{i=0}^{\infty} \frac{d\gamma_i}{ds} \mu^i + [\alpha s + \beta(1 - \alpha)]$$

$$\sum_{i=0}^{\infty} \mu^i \sum_{j=0}^{\infty} \frac{d\gamma_j}{ds} \gamma_{i-j}$$

Reading off terms in each power of μ is now easy:

$$\mu^{-1} : s - \gamma_0(\alpha s + 1 - \alpha) = 0$$

$$\mu^i, i \geq 0 : -\gamma_{i+1}(\alpha s + 1 - \alpha) = -s\frac{d\gamma_i}{ds} + [\alpha s + \beta(1 - \alpha)] \sum_{j=0}^{\infty} \frac{d\gamma_j}{ds} \gamma_{i-j}$$

Equation 3.29 is the first of these equations that we have encountered, along with its solution. Rewriting the second equation in the form

$$\gamma_{i+1} = \frac{1}{\alpha s + 1 - \alpha} \left\{ s\frac{d\gamma_i}{ds} - [\alpha s + \beta(1 - \alpha)] \sum_{j=0}^{\infty} \frac{d\gamma_j}{ds} \gamma_{i-j} \right\}.$$

Keep in mind that all we need to know to figure out the series $i + 1$st term is the formula up to i. For instance, the formula for γ_1 is

$$
\begin{aligned}
\gamma_1 &= \frac{1}{\alpha s + 1 - \alpha} \left\{ s \frac{d\gamma_0}{ds} - [\alpha s + \beta(1 - \alpha)] \frac{d\gamma_0}{ds} \gamma_0 \right\} \\
&= \frac{s(1 - \alpha)^2(1 - \beta)}{(\alpha s + 1 - \alpha)^4}
\end{aligned}
$$

These calculations, as you may guess, quickly get challenging. In the past, we would meticulously simplify perturbation series by hand and fill entire notebooks with them. We frequently spent more time checking our results than we did performing the initial calculation because it was so error-prone. We simply use Maple these days. Even though it's considerably quicker, it's a good idea to manually calculate a few terms to make sure Maple is performing the calculations accurately.

The problems to be covered are what the following section will refer to as layer-type problems. The matching of asymptotic expansions, the domains of validity of such expansions, the overlap, the formal validity, enormous integrated effects of formally small terms, switchback, limit processes, equations, etc are the primary ideas to be examined. In the middle of the 1950s, Saul Kaplun introduced several of these ideas. Older approaches were much better understood as a result of Kaplun's study of the underlying concepts and the introduction of new concepts, which in turn led to an extension of those techniques.

There are numerous perturbation problems that are singular but are not of the layer type. Many of these problems fall within the secular type. The reader is directed to Cole (1968) and Kevorkian (1966), both of which examine these issues in-depth and with numerous examples. The strained coordinates method employed by Whitham and Lighthill is similarly left out of the discussion; Van Dyke (1964) provides references.

3.6   A simple physical example

We will begin with a very simple physical example that demonstrates how both secular-type problems and layer-type problems may occur before attempting any comprehensive discussion of what a single perturbation problem is.

3.6.1 Time scales

A linear spring-mass system with mass m, a damping coefficient of 2β, and a spring constant of k is taken into consideration. then, the equation

$$m\frac{d^2 y_d}{(dt_d)^2} + 2\beta\frac{dy_d}{dt_d} + ky_d = 0 \tag{3.30}$$

where t_d is the dimensional time and y_d is the amplitude expressed in terms of dimensional length. We are aware that non-dimensional terminology may be used to express the response. We could think of non-dimensional amplitude as,

$$y = y_d/L, \tag{3.31}$$

where L is an amplitude (of dimensional length) that is usual. We first notice that there are two independent approaches to forming parameters of dimension time from the initial parameters in order to find a nondimensional time scale. We could define

$$T_1 = \sqrt{m/k} \tag{3.32}$$

which is the inverse of the frequency. Another time parameter is

$$T_2 = \beta/k, \tag{3.33}$$

From these we may form further time parameters

$$T = T_1^a T_2^b, \qquad a + b = 1, \tag{3.34}$$

One way of non-dimensionalizing the time-coordinate is to define

$$t = t_d/T_1. \tag{3.35}$$

Then (3.30) reads

$$\frac{d^2 y}{dt^2} + 2\varepsilon\frac{dy}{dt} + y = 0 \tag{3.36}$$

where

$$\varepsilon = \beta/\sqrt{mk} = T_2/T_1. \tag{3.37}$$

3.6.2 A secular-type problem.

Let's say we now have a "small damping" system. This indicates that when expressed in non-dimensional form,

$$\beta \ll \sqrt{mk} \qquad \text{or, equivalently,} \qquad \varepsilon \ll 1. \tag{3.38}$$

Making assumption (3.38) we study how one may solve (3.36) by perturbation techniques. We take as initial conditions

$$y(0) = 0, \qquad (dy/dt)_{t=0} = 1. \tag{3.39}$$

It seems natural to assume a first approximation y_0 satisfying

$$d^2 y_0/dt^2 + y_0 = 0 \tag{3.40}$$

and the initial condition (3.39) then

$$y_0 = \sin t. \tag{3.41}$$

If we assume that y has an asymptotic expansion

$$y \sim \sum_{k=0}^{\infty} \varepsilon^k y_k(t) \tag{3.42}$$

then y_1 must satisfy

$$d^2 y_1/dt^2 + y_1 = -2\cos t \tag{3.43a}$$

and

$$y_1(0) = y_1'(0) = 0 \tag{3.43b}$$

Hence

$$y_1 = -t \sin t \tag{3.44}$$

and

$$y = y_0 + \varepsilon y_1 + O(\varepsilon^2) = \sin t - \varepsilon t \sin t + O(\varepsilon)^2 \tag{3.45}$$

Now, the question is whether or not (3.45) is a "good approximation." Obviously, the answer varies depending on the time frame at hand. Think about the time frame

$$D = [0, \tau] \tag{3.46}$$

If $\tau = O(1)$, then, as is easily checked, there exist a constant K such that

$$|y - (y_0 + \varepsilon y_1)| \leqq K\varepsilon^2 \tag{3.47}$$

for t in D. When this happens, the error is consistently small and of the formally specified order. But if τ is large, like $\tau = O(1/\varepsilon)$, then (3.47) is no longer true. Using simple language, we state that while $-\varepsilon t \sin t$ has ε as its formal order, unity is the actual order for $t = O(1/\varepsilon)$. Therefore, Yo + ey is not valid on the interval [0, 1/e] but is uniformly valid to order e 2 on the interval $[0, 1]$. The core of a singular perturbation problem is this circumstance: The "simple" perturbation (3.42), however, is not consistently valid.

The approximation can be verified by comparing it to the exact solution.

$$y \approx e^{-\varepsilon t} \sin t \tag{3.48}$$

is good to order unity on the length $1/\varepsilon$ interval. It is obvious that the exponential function's expansion into powers of εt is what causes the term (3.44) to exist. If the order is ε^{-1}, no partial sum of this expansion is valid to order ε. Therefore, it is incorrect to assume that a uniformly valid expansion might have the form (3.42). A term like (3.44) is referred to as a secular term. It happens because of the resonant forcing function in equation (3.43a). However, in the current problem, this resonance is a false mathematical resonance brought on by a poor perturbation approach, not a genuine physical event. We will refer to the problems covered here as secular-type problems. Such problems require a systematic perturbation approach that demonstrates how secular terms can be avoided. Such methods go outside the scope of this book. Kevorkian (1966) and Cole (1968) go into great detail about them. We will just highlight the following points. In (3.48), the $\sin t$ changes with $t = t_d/T_1$, whereas the exponential changes an order of magnitude more slowly, with $\varepsilon t = [(T_2/T_1)(t_d/T_1)] = t_d/T_3$.

$$T_3 = T_1^2 T_2^{-1} = m/\beta \tag{3.49}$$

As a result, we employ the "normal" T_1 and the extremely lengthy T_3, two-time scales with different lengths. In accordance with this, we employ two non-dimensional time variables, t, and εt. The first one is referred to as the "fast" variable, and the second is the "slow" variable.

3.6.3   A layer-type problem

Now, we'll talk about the other extreme, which is a very small mass or very huge damping.

$$mk \ll \beta^2 \tag{3.50}$$

we may write (3.30) as

$$\eta \frac{d^2 y}{ds^2} + 2 \frac{dy}{ds} + y = 0 \tag{3.51}$$

with

$$s = \frac{t_d k}{\beta} = \frac{t_d}{T_2}, \qquad \eta = \frac{km}{\beta^2} = \frac{1}{\varepsilon^2} = \frac{T_1^2}{T_2^2} \tag{3.52}$$

Now, the small parameter is η. As an initial condition, we take

$$y = A \tag{3.53a}$$

$$dy/ds = B \qquad \text{as} \qquad s = 0 \tag{3.53b}$$

If we now consider an expansion of the form

$$y \equiv \sum_{k=0}^{\infty} \eta^k y_k(s) \tag{3.54}$$

Even before we begin to address the issue, we become aware that something is incorrect. The first-order equation would be obeyed by the leading term $y_0(s)$.

$$2\frac{dy_0}{ds} + y_0 = 0 \tag{3.55}$$

As a result, it is impossible to meet both initial conditions (3.53)
. The precise solution reveals that

$$y \approx Ae^{-s/2} - \frac{1}{4}(A + 2B)e^{-2s/n} \cdot \eta \tag{3.56a}$$

$$dy/ds \approx -\frac{1}{2}Ae^{-s/2} + \frac{1}{2}(A + 2B)e^{-2s/n} \tag{3.56b}$$

is a close approximation to order unity for any $[0, s]$ interval. The term of formal order η in y is ignored, and the result is $Ae^{-s/2}$, which satisfies both the initial condition (3.53a) and the condition (3.55). However, it provides the incorrect velocity at s_0 and thus for small values of s as well. To fix this error, a second term is required.

Once more, it is clear that using two-time scales and two non-dimensional time variables is necessary for an accurate approximation.

$$s = t_d/T_2; \qquad T_2 = \beta/k \tag{3.57a}$$

$$s/\eta = t_d/T_3; \qquad T_3 = T_1^2 T_2^{-1} = m/\beta \tag{3.57b}$$

If T_2 is regarded as "normal", T_3 is a relatively short scale. In order to express a universally accurate approximation, both scales must be employed.

The two perturbation issues discussed above share the following same characteristics: The assumption that the expansion in ε is a straightforward power series in ε results in approximations in both instances that are not consistently accurate. The leading term of such a power series is valid to order unity at finite intervals in the first case. With the exception of a narrow layer very close to the origin, where the error in the velocity is of order unity, it is true everywhere in the second case.

We observe that this correction term, which is proportional to $e^{-2s/\eta}/$, changes extremely quickly, with changes of order unity occurring when changes in the variable s occur with order η. In order to find the solution, one needs to combine two-time scales with distinct orders of magnitude. Thus, we apply an *approach of multiple scales* (in this case, two-time scales) to both situations. In layer-type problems, one scale is used in one region (which will later be referred to as the "outer") and the other scale, which is far smaller than the first, is used in another region, which is a very thin layer (which will later be referred to as the "inner").

Both scales are applied simultaneously in problems of the secular type. In our example, a carrier wave called $\sin t$ has an amplitude that progressively changes as $e^{-\varepsilon t}$. While the secular terms were caused by improper perturbation methods, it should be noted that there is a valid secular impact; the damping is only apparent if the system is studied over an extremely long period of time [4]. We see that having two distinct time scales with different orders of magnitude provides one the tiny parameter e, which makes it possible to employ perturbation methods.

The question of which scale to use to non-dimensionalize the time variables arises as a result of the existence of two such independent scales. In reality, we observed that both were required in each situation, i.e., the issues are singular. In general, a perturbation scheme is likely to be singular when the tiny parameter e is a ratio of two lengths (or periods), as was emphasised by Van Dyke (1964, p. 80). Undoubtedly, there are numerous singular perturbation problems that do not fall under this category. One such instance is flow at extremely high Mach numbers, which is extensively explored in the literature (see, for instance, Van Dyke (1964)).

3.7   General concepts in perturbation problems

We will go over a few basic asymptotic expansions principles in this section. The in-depth analysis of particular situations in the following sections will need the use of these principles.

---

[4]The expression "secular" originated in classical celestial mechanics, where a century was considered a long time unit as opposed to short time units such as year or month.

3.7.1   Coordinates and parameters; approximations

In perturbation problems, two sets of complex-or real-valued arguments are taken into account on which the function $f(x_1, \ldots, x_n; \varepsilon_1, \ldots, \varepsilon_n)$ is dependent. One set $(x_1, \ldots, x_n)$ is called "variables, and another set $(\varepsilon_1, \ldots, \varepsilon_n)$ is called parameters. The distinction between variables and parameters depends on the context rather than being a property of the function itself. The distinction may be made based on physical principles; alternatively, if the function is implicitly defined by differential equations and boundary conditions, then the equations only contain derivatives with respect to the variables and none with regard to the parameters. The distinction in a perturbation problem results from how the variables x_j and ε_j are handled. We'll just use one parameter, ε, and one variable, x, for our convenience. In a perturbation problem, when ε is very close to a notable value that we may assume to be zero without losing generality, we look for approximations to $y(x, \varepsilon)$, which should be uniformly valid for all x in a closed domain D. If $f(x, \varepsilon)$ approximates $y(x, \varepsilon)$ uniformly valid to order $\zeta(\varepsilon)$, then we say that,

$$\lim_{\varepsilon \to \infty} \frac{y(x, \varepsilon) - f(x, \varepsilon)}{\zeta(\varepsilon)} = 0 \text{ uniformly for } x \text{ in } D \qquad (3.58)$$

Naturally, there is no differentiation in this definition between a function that is an approximation and one that is "the real thing." The function $y(x, \varepsilon)$ in a perturbation problem is typically implicitly specified by a number of equations (differential, integral, functional, boundary conditions). The goal of perturbation methods is to create approximate functions $f(x, \varepsilon)$, typically through the analysis of simplified equations. A series of approximations are frequently built, as will be detailed below.

Only nonnegative real values of ε may be a parameter in a physical problem such that they have physical significance. It is well known that it is preferable to let ε assume complex values in order to conduct a complete mathematical analysis of $y(x, \varepsilon)$. Cases, where $y(x, \varepsilon)$ possesses an essential singularity at $\varepsilon = 0$ in the complex ε-plane, will be dealt with frequently. In this situation, the path through which e approaches 0 may have an impact on the limit in (3.58). Unless otherwise stated, we will assume that ε approaches zero through positive numbers in the paragraphs that follow. In some wedge of the complex ε-plane surrounding these values, the results are typically valid. However, we won't go into detail about complex values.

3.7.2   Sequence of approximations.

In order to solve a perturbation problem, one typically builds what is, in theory, the asymptotic expansion of $y(x, \varepsilon)$ at $\varepsilon = 0$.

Let $\zeta_i(\varepsilon), j = 0, 1, \ldots$, be an asymptotic sequence (for $\varepsilon \downarrow 0$), i.e.[5],

$$\lim_{\varepsilon \downarrow 0} \zeta_{j+1}(\varepsilon)/\zeta_j(\varepsilon) = 0 \qquad (3.59)$$

A sequence of functions $a_0(x, \varepsilon), a_1(x, \varepsilon), \ldots$ such that for any j, is a sequence of uniformly valid asymptotic approximations to $y(x, \varepsilon)$,

$$\lim_{\varepsilon \downarrow 0} \frac{y(x, \varepsilon) - a_j(x, \varepsilon)}{\zeta(\varepsilon)} = 0 \qquad (3.60)$$

uniformly in D. As a result, any $a_j(x, \varepsilon)$ is a uniformly valid approximation to order. The gauge functions for this approximation method are referred to as the $\zeta_j(\varepsilon)$. Finding a sequence $b_j(x, \varepsilon)$ as given below to construct the $a_j(x, \varepsilon)$ is a very popular technique.

$$a_j(x, \varepsilon) = \sum_{k=0}^{j} b_k(x, \varepsilon). \qquad (3.61)$$

We claim that the formal sum $\sum_{k=0}^{\infty} b_k(x, \varepsilon)$ is an asymptotic expansion of $y(x, \varepsilon)$ relative to the gauge functions $\zeta_j(\varepsilon)$, uniformly valid in D, if (3.60) and (3.61) hold, and write

$$y(x, \varepsilon) \simeq \sum_{k=0}^{\infty} b_k(x, \varepsilon). \qquad (3.62)$$

The reader is directed to traditional literature, such as Wasow (1965) or Erdelyi (1956), for information on the fundamental characteristics of asymptotic expansions.

In conclusion, a perturbation method as used here is a technique for creating an asymptotic expansion, or the start of one, of a function $y(x, \varepsilon)$. There will be approximations to $y(x, \varepsilon)$. in the partial sums of (3.62). The error term is an order estimate, though. We may change (3.60) to

$$y(x, \varepsilon) - a_j(x, \varepsilon) = o(\zeta_j(\varepsilon)). \qquad (3.63)$$

---

[5]To express relations such as (3.59) we shall often use the simpler notation $\zeta_{j+1} \ll \zeta_j$ or $\zeta_{j+1} = O(\zeta_j)$.

Finding functions $\zeta_j^*(\varepsilon)$ where $\zeta_j^* = o(\zeta_j)$, such that $|y(x,\varepsilon) - a_j(x,\varepsilon)| \leqq \zeta_j^*(\varepsilon)$ for all x in D is required to translate this into a numerical error estimate. We won't discuss the approximation scheme's numerical aspect here.

3.7.3 Regular and singular perturbations.

It might be tempting to categorize perturbation problems as regular or singular depending on whether $y(x,\varepsilon)$ is regular or singular around $\varepsilon = 0$. However, this results in numerous complications. The following is one of them: If x^* and ε^* by [6]

$$x^* = \text{function of } x \text{ and } \varepsilon \tag{3.64a}$$
$$\varepsilon^* = \text{function of } \varepsilon \text{ such that } \varepsilon^* = 0 \text{ for } \varepsilon = 0, \tag{3.64b}$$
$$f^*(x^*,\varepsilon^*) = f(x,\varepsilon), \tag{3.64c}$$

then in the sense described above, ε^* is still a parameter and x^* is still a variable. It is possible for f to be singular at $\varepsilon = 0$ but f^* is regular at $\varepsilon^* = 0$. A simple illustration is, with

$$D = \{x|x0\}, \tag{3.65a}$$
$$f(x,\varepsilon) = e^{-x/\varepsilon} + x \tag{3.65b}$$
$$x^* = x/\varepsilon, \qquad \varepsilon^* = \varepsilon, \tag{3.65c}$$
$$f^*(x^*,\varepsilon^*) = \varepsilon^{-x^*} + \varepsilon x^*. \tag{3.65d}$$

With regard to the domain D and the gauge functions $\zeta_j(\varepsilon)$, we claim that a uniformly valid asymptotic expansion (3.62) is regular if each $b_j(x,\varepsilon)$ can be expressed as

$$b_j = \beta_j(\varepsilon)b_j(x), \tag{3.66}$$

that is, if (3.62) has the form

$$y(x,\varepsilon) \simeq \sum_{j=0}^{\infty} \beta_j(\varepsilon)b_j(x), \tag{3.67}$$

where the sequence of gauge functions $\zeta_j(\varepsilon)$ may or may not coincide with the asymptotic sequence formed by the $\beta_j(\varepsilon)$. The $\beta_j(\varepsilon)$ need not be powers of ε, but we do allow, for example,

---

[6]We shall denote the transformation (3.64) by the vague term "coordinate change."

$$\beta_0 = 1, \qquad \beta_1 = \varepsilon In\varepsilon, \qquad \beta_2 = \varepsilon, \qquad etc$$

Therefore, it is not required that $f(x,\varepsilon)$ be regular at $\varepsilon = 0$. It is clear that the two terms on the right-hand side of (3.65b) do not make up a regular expansion. Less plainly, since the term εx^* is not uniformly small in the domain D under consideration, if $\zeta_j = \beta_j = \varepsilon^j$, then (3.65d) is also not a regular expansion. Again, it becomes clear that the existence of a regular expansion does not rely on the regularity of ε at $\varepsilon = 0$.

The construction of the $\beta_j(\varepsilon)$ and $b_j(x)$ using a regular perturbation method is methodical and could possibly be preceded by a change in coordinates of the kind 3.64.

A general definition of a singular perturbation technique is a non-regular method. In layer-type problems, we generally build expansions of the kind (3.67) that are not uniformly valid and must be complemented by one or more expansions in order to create expansions that are uniformly valid.

$$y(x,\varepsilon) \simeq \sum_{j=0}^{\infty} \beta_j(\varepsilon)a_j(x^*), \tag{3.68}$$

where x^* is a new variable of a different magnitude than x, and function of ε such as $x^* = x/\varepsilon,$. The form is identical to that of (3.67), with the exception that we use the coordinate x*. In fact, as we'll see further, one or more of the expansions may not be of the special form (3.67), (3.68) but will still be of the general form (3.62). It follows that the various expansions must, in some way, provide approximations that are consistently valid in D. The following section will go over this crucial idea.

Strictly speaking, one should talk about singular perturbation approaches or procedures rather than singular perturbation *problems*. Consider a differential equation as an example with boundary conditions that are so obvious and straightforward to solve using Fourier analysis. However, we do not refer to this problem as a Fourier problem. Similarly, there are perturbation problems for which constructing a singular asymptotic expansion is natural. A clever a priori coordinate change, however, might have made it possible to build a regular asymptotic expansion. A regular expansion can be built using the singular expansion as a foundation. However, we will continue to refer to this problem as a "singular perturbation problem" throughout this study

3.7.4   Validity of an approximation

Obtaining approximations that are consistently valid in the specified domain D is the goal of perturbation methods. The way one works with singular perturbation techniques (of the layer kind) is common with approximations that are only accurate in restricted domains. For these approximations, the idea of a uniform domain of validity must be defined. As a result, we will generalise (3.58).

$$y(x, \varepsilon) = e^{-x/\varepsilon} + x + \varepsilon \tag{3.69a}$$

$$f(x, \varepsilon) = x + \varepsilon \tag{3.69b}$$

$$g(x, \varepsilon) = e^{-x/\varepsilon} \tag{3.69c}$$

We will see $y(x, \varepsilon)$ as "the exact solution" while approximating $f(x, \varepsilon)$ and $g(x, \varepsilon)$. The closed interval $[0, 1]$ will be the variable x's domain. We can see that f is a close approximation to yy and is not valid on $[0, 1]$ (and is not uniformly valid on $[0, 1]$ even if one redefines $f(0, \varepsilon)$ to be $= 1$). Instead, it is uniformly valid to order unity on any interval $[x_0, 1]$ provided $0 < x_0 \leqq 1$.

The discussion of the situation where $f(x, \varepsilon)$, is a uniformly valid approximation seems to be exhausted with the remarks just made. But expanding the idea of "uniform validity" is crucial for the theory of matching. Think about the continuum $[\sqrt{\varepsilon}, 1]$ where ε ranges in some open interval $0 < \varepsilon < \varepsilon_0 \leqq 1$. The left endpoint of any such period has the maximum error, which is equal to $e^{-1/\sqrt{\varepsilon}}$. Then it becomes evident that for any value of $\delta > 0$ one can find an ε_δ such that $|y - f| < \delta$ for all $\varepsilon, 0 < \varepsilon < \varepsilon_\delta$ and all x in $[\sqrt{\varepsilon}, 1]$ that is, in the vertically striped region of the (x, ε)-plane as shown in the figure 3.1. The corresponding domain for an interval with the horizontally striped rectangle in the figure 3.1 represents the corresponding domain for an interval with fixed endpoints. Instead of referring to $[\sqrt{\varepsilon}, 1]$ as a continuum of intervals, we will simply refer to it as an interval with a moving endpoint. It's reasonable to state that f is a uniformly valid approximation to y on the range $[\sqrt{\varepsilon}, 1]$. We see that in this sentence, $\sqrt{\varepsilon}$ may be replaced by any function η such that $0 < \eta(\varepsilon) < 1$ for ε in some interval $(0, \varepsilon_0)$ provided that $\varepsilon \ll \eta$, i.e., $\lim(\varepsilon/\eta) = 0$ as $\varepsilon \downarrow 0$. Similar to this, we observe that $g(x, \varepsilon)$ is an approximation to $y(x, \varepsilon)$ uniformly valid on any interval $[0, \eta]$ provided $\eta(\varepsilon) \ll 1$, and $0 < \eta(\varepsilon) < 1$ for ε sufficiently small.

The following definitions are suggested by the examples above.

Definition 3.7.1. *For any two $v(\varepsilon)$ and $\mu(\varepsilon)$) in $\mathfrak{D}$, a function $w(x, \varepsilon)$*

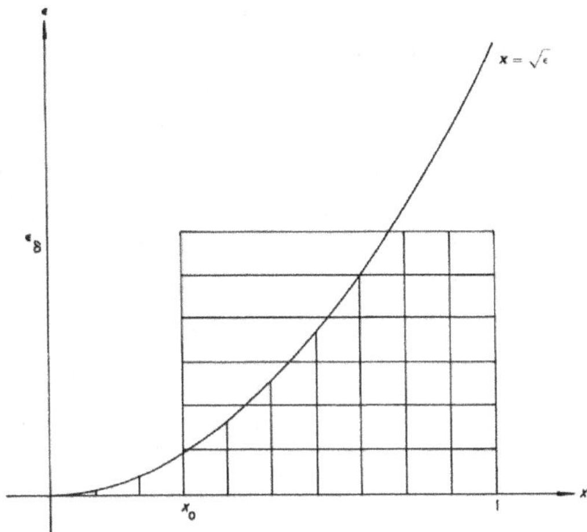

Figure 3.1: Domains of validity

converges uniformly to zero with respect to $\mathfrak{D}$ (a class of functions $\eta(\varepsilon)$) if it does so uniformly on the interval $[\mu, v]$. For each $\delta > 0$, an ε_δ exists such that $|w(x, \varepsilon)| < \delta$ for $\varepsilon < \varepsilon_\delta$ and $\mu(\varepsilon) \leqq x \leqq v(\varepsilon)$.

Although there are no problems in practice, this definition still needs some refinement. The interval $[\mu, v]$ is empty, for values of ε satisfying the condition that $v(\varepsilon) < \mu(\varepsilon)$. Saying that the error is zero on an empty interval is the most feasible way to handle this situation (see Kaplun (1967)). Additionally, the range $[\mu, v]$ may partially fall beyond the range of the definition of $w(x, \varepsilon)$. By properly reducing the functions, this may be dealt with easily. The following is more important. There will be exceptional situations in layer-type problems where the tools we are creating here will be required. Each of these points can always have a coordinate introduced into its study so that it is at the origin. Additionally, for such points, the behavior of the function on one side may differ significantly from that on the other ($e^{-x/\varepsilon}$ is a well-known example). We may use a coordinate that is positive on one side to study the behavior on that side. We will therefore presume that functions $\eta(\varepsilon)$ are non-negative. More specifically, we'll suppose that either $\eta(\varepsilon) > 0$ or $\eta(\varepsilon) = 0$ for the entire range $(0, \varepsilon_0)$. Although the motivation for these principles is not yet clear, the different instances that will be provided later will help to clarify them. With these limitations on our function $\eta(\varepsilon)$, we define it as follows.

Definition 3.7.2. *If $y(x,\varepsilon) - f(x,\varepsilon))/\zeta(\varepsilon)$ converges to zero uniformly with respect to $\mathfrak{D}$, then a function $f(x,\varepsilon)$ is an approximation to the function $y(x,\varepsilon)$ uniformly valid to order $\zeta(\varepsilon)$ with respect to the function class $\mathfrak{D}$.*

The idea of uniform convergence on an interval is thus generalized to include uniform convergence on a function class in (3.7.1). The concept of uniform validity on an interval is similarly generalized to include uniform validity on a function class in (3.7.2). [7]

3.7.5   Domains of validity

Returning to the example from (3.69), we can see that the function class is the domain of uniform validity (to order unity) of $f(x,\varepsilon)$ as an approximation of $y(x,\varepsilon)$.

$$\mathfrak{D} = \{\eta | \varepsilon \ll \eta\} \tag{3.70a}$$

Similarly the domain of validity for the approximation $g(x,\varepsilon)$ is

$$\mathfrak{D} = \{\eta | \eta \ll 1\} \tag{3.70b}$$

As mentioned previously, the function $\eta(\varepsilon)$ is constrained in this case.

The next lemma is trivial.

Lemma 3.7.1. *Let $\mathfrak{D}$ be a domain of validity of an approximation $f(x,\varepsilon)$ and let $\eta_1 < \eta_3 < \eta_2$ with η_1 and η_2 in $\mathfrak{D}$. Then the domain of validity can be extended to include η_3. The same conclusion follows if $\eta_1 \ll \eta_3 < \eta_2$*

Thus, adding intermediate functions to a domain of validity is always possible.

There will typically be a maximal domain of validity in practical situations. When we refer to the domain of validity, we mean the maximal one.

3.7.6   Order classes

In our example, we observe that the domain to which a function $\eta(\varepsilon)$ belongs (3.70a) or (3.70b) simply depends on the function's order of magnitude. The following definition makes it easy to formalize this.

---

[7]The definition of asymptotic expansion introduced earlier is a natural but trivial generalization of Poincare's definition. On the other hand, the introduction of the concepts used in (3.7.1) and (3.7.2), due to Kaplun, is definitely nontrivial these concepts will be basic tools in the study of the nature of matching.

Definition 3.7.3. *If* $\lim \eta/\zeta$ *as* $\varepsilon \downarrow 0$ *exists and is distinct from zero and infinity, the functions* η *and* ζ *are o-equivalent.*

This clearly represents an equivalence relation. Thus, the equivalence class or order class (o-class) may be defined.

$$\mathrm{ord}\ \eta = \{\zeta | \zeta \ \text{o-equivalent to}\ \eta\}.$$

We may also define the following more inclusive relation.

Definition 3.7.4. *If there are constants* A *and* B, $0 < A < B$, *such that* $A < \zeta/\eta < B$ *for sufficiently small* ε , *then the functions* η *and* ζ *are O-equivalent.*

The O-class is similarly defined.

$$\mathrm{ord}\ \eta = \{\zeta | \zeta \ \text{O-equivalent to}\ \eta\}.$$

There is a natural way of introducing a partial ordering among the equivalence classes:

$$\mathrm{ord}\ \eta < \mathrm{ord}\ \zeta \ \text{and}\ \mathrm{Ord}\ \eta < \mathrm{Ord}\ \zeta \qquad (3.71a)$$

both mean

$$\lim \eta/\zeta = 0, \qquad i.e. \eta \ll \zeta. \qquad (3.71b)$$

The ordering is just "partial" there may be equivalence classes that are incompatible, i.e. distinct yet neither greater than the other. Since ord $\eta_1 < \mathrm{ord}\ \eta_2$ implies Ord $\eta_1 < \mathrm{Ord}\ \eta_2$, O- classes that are pairwise incompatible make up the class Ord η .

Any rigorous theory of singular perturbations will likely place a significant emphasis on the idea of order classes. Additionally, the structure of sets of these classes gives rise to a number of intriguing problems. The reader is referred to Kaplun (1967) for additional details.

3.7.7   Overlap

The half-infinite intervals that make up the domains in (3.70a) and (3.70b) are obviously open. The domains of validity have the intersection

$$\mathfrak{D}_f \cap \mathfrak{D}_g = \{\eta | \varepsilon \ll \eta \ll 1\}.$$

Overlap domain of the two approximations (3.69b) and (3.69c) is referred to as this. To be more specific, it is the overlap domain to *order unity*.

Together, they simultaneously *cover* the entire interval under discussion. $\mathfrak{D}_g$ covers all "infinitesimal" x values, i.e., all $x = x_\eta \cdot \eta, \eta = O(1)$, as well as the function $\eta \cong 0$. All constant functions $\eta = a, 0 < a \leq 1$, and enough infinitesimals to overlap with $\mathfrak{D}_g$ are all included in $\mathfrak{D}_f$. The two domains' union is thus referred to as the complete interval.

$$\mathfrak{D}_f \cup \mathfrak{D}_g = [0, 1].$$

The expression that the functions f and g of (3.69b) and (3.69c) combined approximate the function y of (3.69a), which is uniformly valid to order unity on the interval $[0, 1]$ will be used.

3.7.8   Limit processes

Limit processes are an essential tool in singular perturbation techniques. The η-limit of a function $y(x, \varepsilon)$ is defined as follows. We begin by introducing the variable x_n

$$\eta x_n = x. \tag{3.72}$$

Then the η-limit of f is

$$\lim_{\eta} f = \lim f \quad as \quad \varepsilon \downarrow 0 \quad with \quad x_\eta \quad \text{fixed.} \tag{3.73}$$

We can therefore rewrite f as a function of x_η, and ε after ensuring that it has the required continuity and then put $\varepsilon = 0$. Take into account the definition of $y(x, \varepsilon)$ in (3.69a) to demonstrate this idea. Then

$$\lim_{\eta} y = e^{-x/\varepsilon} \quad for \quad \eta = \varepsilon \tag{3.74a}$$

$$\lim_{\eta} y = 0 \quad for \quad \eta = \sqrt{\varepsilon} \tag{3.74b}$$

$$\lim_{\eta} y = x \quad for \quad \eta = 1 \tag{3.74c}$$

3.7.9   A fundamental lemma

The proof of the following lemma will be omitted because it is trivial. However, it is crucial to the theory of matching (cf. (3.7.1)).

Lemma 3.7.2. *Let $\mathfrak{D}$ be an overlap domain to order $\zeta(\varepsilon)$ and let $f(x, \varepsilon)$ and $g(x, \varepsilon)$ be approximations to $y(x, \varepsilon)$. Assume that $\eta_1 \ll \eta \ll \eta_2$ for η_1 and η_2 be in $\mathfrak{D}$. Then*

$$\lim_{\eta} \frac{f(x, \varepsilon) - g(x, \varepsilon)}{\zeta(\varepsilon)} = 0$$

3.7.10   Matching

Let f and g represent two asymptotic approximations of y. If their respective domains of validity overlap, they are said to *match*. For instance, if f and g have been derived via approximation equations but g still involves one unknown constant, then one may find it by matching, or by comparing in the domains of overlap between them.

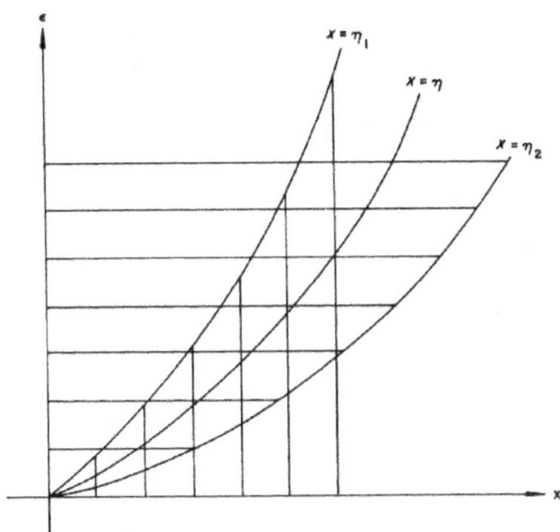

Figure 3.2: Matching in overlap domain

In *principle*[8] , the matching might be done by utilizing (3.7.2). Taking the $\lim_\eta$ entails allowing ε to get closer to zero on the curve $x = \eta(\varepsilon)$ as shown in Fig.3.2 . Here, vertical stripes denote a portion of the domain of validity of f, and horizontal stripes denote the domain of validity of g.

3.7.11   The extension theorem

It is obvious that in order to employ matching, we need criteria to determine as much of the approximation's valid domain as possible. For the cases in examples discussed earlier The domains of validity will be determined in a significant way by the following theorem, which is attributed to Kaplun.

Theorem 3.7.3. *Let $w(x,\varepsilon)$ tend to zero uniformly with respect to a class of functions $\mathfrak{D}$ which is maximal and let all functions of an o-class Y be in $\mathfrak{D}$. Then there exists an o-class X in $\mathfrak{D}$ such that $X < Y$.*

---

[8]How it is done in practice will be discussed later.

Proof. First assuming $Y = \text{ord}1$. Consider the interval $0 \leq x \leq 1$ in order to correct the ideas. By making the assumption that $c > 0$ and $\delta > 0$ are constants, it is possible to discover an $\varepsilon_\delta > 0$ such that $w(x, \varepsilon) < \delta$ for $c \leq x \leq 1$, $0 < \varepsilon \leq \varepsilon_\delta$. Using this characteristic for a sequence of constants c_n, and a sequence δ_n that both tend to zero is the key to the proof's trick. For example, one may enter, $\delta_n = c_n = 1/n$. We can choose an ε_n, such that $w < 1/n$ for $1/n \leq x \leq 1$; $0 < \varepsilon \leq \varepsilon_n$ as all constants $1/n$ belong to ord1.

Assuming that η is a linear function of ε between any two adjacent values ε_k and ε_{k+1}, we now define a function $\eta(\varepsilon)$ by the requirement $\eta(\varepsilon_{n+1}) = 1/n$. Now that we know an X, we find ord $\eta < X < Y$. Anyone may observe that any $\mathfrak{D}$ contains such X. In the general instance, we replace x by x/ζ in the previous argument when $Y = \text{Ord } \zeta$. Clearly, the theorem may be demonstrated by changing "$<$" to "$>$". The proof also demonstrates that all of the 0-class, which corresponds to X is in $\mathfrak{D}$. $\square$

A maximal domain of validity cannot have an infimum or a supremum, according to the extension theorem. It is open in the sense that there are 0-classes X and Z in the domain for any 0-class Y, such that $X < Y < Z$. Now, using numerous model examples, we'll illustrate the singular perturbations technique and the underlying theories and heuristics.

3.8   First model equation

3.8.1   Formulation of problem

We use a function $y(x, \varepsilon)$ that is implicitly defined by the differential equation as Example I

$$\varepsilon \frac{d^2 y}{dx^2} + \frac{dy}{dx} - a - 2bx = 0 \tag{3.75a}$$

and the boundary conditions

$$y(0, \varepsilon) = 0 \tag{3.75b}$$

$$y(1, \varepsilon) = 1 \tag{3.75c}$$

We obtain the explicit expression by solving the aforementioned equation. Here a and b are constants, independent of ε

$$y = (1-a-b+2\varepsilon b)\left(\frac{1-e^{-x/\varepsilon}}{1-e^{-1/\varepsilon}}\right)+bx^2+ax-2\varepsilon bx = \tilde{y}+O(e^{-1/\varepsilon}) \quad (3.76)$$

where $\tilde{y}$ is obtained from y by leaving out the expression $e^{-1/\varepsilon}$ in the y's first term's denominator.

We will now demonstrate how a method of singular perturbations applied to (3.75a), (3.75b), (3.75c) can be used to generate an asymptotic expansion of y. Analysing the outcomes requires comparison with (3.76). As ε trends towards zero through positive real numbers, we will look for an asymptotic expansion that is uniformly valid on the interval $[0, 1]$. We will utilise the sequence $(\varepsilon^k), k = 0, 1, \ldots$, as gauge functions. If a function $\zeta(\varepsilon)$ has a smaller order of magnitude than all gauge functions, it is said to be transcendentally small. If this is true, the function $e^{-1/\varepsilon}$ is transcendentally small in comparison to our gauge functions. We cannot anticipate that our structure will be able to distinguish between y and $\tilde{y}$ in (3.76) using simply the gauge functions.

3.8.2 Perturbation methods

We would presume that $y(x, \varepsilon)$ has an asymptotic expansion in the regular perturbation approach

$$y \simeq \sum_{k=0}^{\infty} \varepsilon^k f_k(x) \quad (3.77)$$

(The ε^k may need to be changed with less simple ε functions, of course.) One obtains an equation and boundary conditions for f_0 by inserting (3.77) into the equation and boundary conditions for y and grouping terms of the same order in ε. Then one finds an equation and boundary conditions for f_1, which may presume that f_0 is known, etc. In this instance, the formula for f_0 is

$$df_0/dx - a - 2bx = 0 \quad (3.78a)$$

This is a first-order equation. Therefore, it will not be able to satisfy both (3.75b) and (3.75c) of the boundary conditions. We impose the second boundary requirement for reasons that will be described later.

$$f_0(1) = 1 \quad (3.78b)$$

This determines f_0 to be

$$f_0 = 1 - a - b + ax + bx^2 \quad (3.79)$$

The value of f_0 at $x = 0$ is $1 - a - b$. We discover that $f_0(0) \neq 0$ when the incidental case of $a + b = 1$ is excluded. The error at $x = 0$ is therefore of order unity even though f_0 is, in some sense to be determined, an approximation to the exact answer y. f_0 is therefore not consistently valid close to $x = 0$. Therefore, we have a problem with a single perturbation. We must complement f_0 to an approximate valid near $x = 0$ in order to obtain an approximation that is uniformly valid. We'll now demonstrate how to build such an approximation. The method's justification will come next.

We provide the expanded variable x^*, which is defined by

$$\varepsilon x^* = x \tag{3.80}$$

Expressed in terms of this variable, (3.75a) becomes

$$\frac{d^2 y}{dx^{*2}} + \frac{dy}{dx^*} - \varepsilon a - 2\varepsilon^2 b x^* = 0 \tag{3.81}$$

This variable's introduction is predicated on the idea that, except from very close to the origin, $f_0(x)$ is a good approximation. More specifically, a layer of thickness ε is present when y shifts from a value of 0 at $x = 0$ to values that are close to $f_0(X)$. The term dy/dx actually has an order of $1/\varepsilon$ rather than its "formal" order of unity due to this quick change. Since $dy/dx = (1/\varepsilon)dy/dx^*$, the introduction of the variable formalises this order determination. We now assume that the approximation to $y(x, \varepsilon)$ has an asymptotic expansion of the following form and is valid close to the origin.

$$y \simeq g_0(x^*) + \varepsilon g_1(x^*) + \dots \tag{3.82}$$

From (3.81) we see that g_0 must satisfy the equation

$$\frac{d^2 g_0}{dx^{*2}} + \frac{dg_0}{dx^*} = 0 \tag{3.83a}$$

Since the boundary condition at $x = 0$ must be met for this approximation to be valid close to the origin, that is,

$$g_0(0) = 0 \tag{3.83b}$$

For the purpose of calculating f_0, we could only apply one boundary condition, which we had to relinquish at $x = 0$. We now have the issue of having very few border conditions: Two boundary conditions must be imposed for the second order equation (3.83a) , and while imposing (3.83a) is natural, the condition at $x = 1$ would be not make sense since g_0 is only meant to be an approximation for x near the origin.

3.8.3 Formal matching

We are currently using a singular perturbations technique that utilises two scales. By applying a scale of order unity (relative to the variable x), one can achieve the approximation $f_0(x)$. Using a scale of order ε, we can derive the approximation $g_0(x^*)$. This scale is only utilised near the value $x = 0$. At the origin, we say there is a boundary layer (or layer of rapid transition) of thickness ε. The *boundary-layer solution* $g_0(x^*)$ is used here since the solution $f_0(x)$ is valid in this layer. We have a layer-type problem since the scale ε is only used in this narrow region. The fact that the missing boundary condition for the boundary-layer solution is provided by *matching* with the solution $f_0(x)$ is a key component of a layer type singular perturbation technique. In this instance, we might make use of a formal matching condition.

$$g_0(\infty) = f_0(0) \equiv 1 - a - b \tag{3.83c}$$

This condition can come as a big surprise. For instance, even though g_0 is only meant to make sense for small x, we use it in (3.83c) at $x^* = \infty$, which would also mean $x = \infty$ if ε were fixed. At the same time, we compare the value of $g_0(\infty)$ with that of $f_0(x)$ for $x = 0$. Currently, all we have is the vague justification that for very small values of ε, there are values of x close to zero where x/ε is close to infinity and where both f_0 and g_0 are valid approximations. The analysis and formalization of this justification is one of the key goals of the current study. This will result in a general way of matching, which will imply (3.83c) in the current particular instance. Using the boundary condition (3.83b) we find

$$g_0 = C(1 - e^{-x^*}) \tag{3.84a}$$

The matching condition (3.83c) then gives

$$C = 1 - a - b \tag{3.84b}$$

3.8.4 Higher order approximations

The expansion (3.77) cannot be expected to be valid at the origin, as we know. We call it an outer expansion. The expansion (3.82), which is anticipated to be valid close to the origin, is referred to as the inner expansion [9] . The first terms of each expansion have been established,

---

[9]The terms "outer" and "inner" are used in this study to refer to expansions, limits, variables, and regions. The names "outer" and "inner" are used for historical reasons. Its roots can be found in the well-known boundary-layer issue of viscous flow past an

and the higher-order terms will now be discovered. We conclude that f_1 satisfies from (3.77),

$$\frac{df_1}{dx} = -\frac{d^2 f_0}{dx^2} = -2b \tag{3.85a}$$

Again we require that $f_0 + \varepsilon f_1$ satisfy the boundary condition at $x = 1$ which gives

$$f_1(1) = 0 \tag{3.85b}$$

and hence

$$f_1 = 2b - 2bx \tag{3.86}$$

From (3.81) we find that g_1 satisfies the equation

$$\frac{d^2 g_1}{dx^{*2}} + \frac{dg_1}{dx^*} - a = -0 \tag{3.87a}$$

The boundary condition at $x = 0$ gives

$$g_1(0) = 0 \tag{3.87b}$$

Hence,

$$g_1 = C_1(1 - e^{-x^*}) + ax^*, \tag{3.88}$$

where C_1 is a constant that can be determined by matching. The straight-forward matching rule (3.83c) is obviously invalid. We replace it with the following rule, which is more expansive but is also a generalization of the justification provided for (3.83c)

$f_0 + \varepsilon f_1$ for x small must agree to order ε with $g_0 + \varepsilon g_1$ for x^* large
$$\tag{3.89}$$

For x^* large the exponential terms are negligible and

$$g_0 + \varepsilon g_1 \simeq 1 - (a + b) + \varepsilon(C_1 + ax^*) \tag{3.90a}$$

For x small we have

$$g_0 + \varepsilon g_1 \simeq ax + 1 - (a + b) + 2b\varepsilon \tag{3.90b}$$

---

object. Actually, the nomenclature used to describe the flow within a domain, which was previously explained, is erroneous. The outer zone is encircled by the inner region in this instance. But we'll stick to the traditional usage. As a result, in our example of order ε, the inner region will always be small in comparison to the outside region. We will give an equivalent way to refer to the outer limit as the first principal limit in the discussion that follows.

Here, we have ignored the second-order terms in x and ε, $bx^2 - 2\varepsilon bx$. If we set $C_1 = 2b$, (3.90a) and (3.90b) agree.

$$g_1 = 2b(1 - e^{-x^*}) + ax^* \tag{3.91}$$

Continuing the same reasoning we find

$$f_k = 0, \qquad k > 1, \tag{3.92}$$

also,

$$g_2 = bx^{*2} - 2bx^* \tag{3.93a}$$

$$g_k = 0 \qquad k > 2 \tag{3.93b}$$

3.8.5 Comparison with exact solution

We have obtained all of the terms for the inner and outer expansion by asymptotic construction. What kind of asymptotic approximation to the precise solution have we found? We require the domain of validity idea introduced in §7 to answer this. First, we look at f_0. Allow x to fall within the range $[\eta, 1]$, where η is a function of ε. Since

$$|y - f_0| = |1 - a - b|e^{-x/\varepsilon} + \varepsilon. \text{ bounded function,}$$

the maximal error occurs at the left endpoint where $x = \eta$. The error tends uniformly to zero in this interval provided $\varepsilon \ll \eta$. Thus, the function f_0 is an approximation to y uniformly valid to order unity on the domain

$$\mathfrak{D} = \{\eta | \varepsilon \ll \eta\} \tag{3.94a}$$

Similarly,

$$f_0 + \varepsilon f_1 \text{ is uniformly valid to all orders } \varepsilon^k \text{ on } \mathfrak{D}_1 \tag{3.94b}$$

It is obvious that the most important attribute is validity to order ε. A straightforward adjustment, such as adding a term cx^2 into the equation (3.75a) or changing $y(1, \varepsilon) = 1 + \varepsilon^2$, would ruin the validity to higher orders.

The inner expansion is valid in the following ways.

Domain of validity of g_0 to order unity is

$$\mathfrak{D}_2 = \{\eta | \eta \ll 1\} \tag{3.95a}$$

Domain of validity of $g_0 + \varepsilon g_1$ to order ε is

$$\mathfrak{D}_3 = \{\eta | \eta \ll \sqrt{\varepsilon}\} \tag{3.95b}$$

The reason for the smaller domain in the second case is that

$$|(y - g_0 - \varepsilon g_1)\varepsilon^{-1}| = bx^2/\varepsilon - 2bx.$$

Finally, $g_0 + \varepsilon g_1 + \varepsilon^2 g_2$ is the exact solution, neglecting the transcendentally small term $e^{-1/\varepsilon}$. This is clearly an accident and would again be spoiled by trivial modifications of the problem. The domains of overlap are, to order unity and ε, respectively:

$$\text{for } f_0 \text{ and } g_0 : \quad \mathfrak{D}_1 \cap \mathfrak{D}_2 = \{\eta | \varepsilon \ll \eta \ll 1\} \tag{3.95c}$$

$$\text{for } f_0 + \varepsilon f_1 \text{ and } g_0 + \varepsilon g_1 : \quad \mathfrak{D}_1 \cap \mathfrak{D}_2 = \{\eta | \varepsilon \ll \eta \ll \sqrt{\varepsilon}\} \tag{3.95d}$$

3.8.6   Matching reconsidered

The success of the matching techniques utilized in the earlier construction is now apparent posterior, consider first f_0 and g_0 . The domain of their overlap is given by (3.95c). For any η in this domain,

$$\lim_{\eta}(f_0 - g_0) = 0 \tag{3.96}$$

in accordance with the fundamental lemma (3.7.2). Any limit with η in the overlap domain has a small $x = \eta x_\eta$ and a large $x^* = \eta x_\eta/\varepsilon$. To be more precise,

$$\lim_{\eta} x = 0 \quad \quad for \, \eta \ll 1, \tag{3.97a}$$

$$\lim_{\eta} x^* = 0 \quad \quad for \, \varepsilon \ll \eta, \tag{3.97b}$$

It follows that, for η in the overlap domain

$$\lim_{\eta} f_0 = f_0(0), \tag{3.98a}$$

$$\lim_{\eta} g_0 = g_0(\infty). \tag{3.98b}$$

(3.96) implies (3.83c)) since the numbers provided by the right sides above are finite.

Similarly,

$$(f_0 + \varepsilon f_1) - (g_0 + \varepsilon g_1) = bx^2 - \varepsilon 2bx + (1 - a - b)e^{-x^*} + \varepsilon 2be^{-x^*}$$

The $\lim_\eta$ of the exponential terms divided by ε vanishes when ε falls within the overlap domain (3.95d). Furthermore,

$$\lim_\eta [(b\eta^2 x_n^2/\varepsilon) - 2b\eta x_n] = 0$$

since $\eta \ll \sqrt{\varepsilon}$. Hence,

$$\lim_\eta \frac{(f_0 + \varepsilon f_1) - (g_0 + \varepsilon g_1)}{den}(\varepsilon) = 0 \qquad (3.99)$$

This demonstrates that the matching result achieved with (3.89) was accurate.[10]

We have seen that the use of Lemma (3.7.2) may allow the formal matching rules (3.83c) and (3.89) to be justified. Lemma matching, as mentioned in the discussion, is by definition a comparison of two approximations in their domain of overlap.[11] On the basis of this, it is possible to create formal rules for matching that do not explicitly include overlap. These rules might, however, only be partially valid, as is unquestionably the case with (3.83c).

However, once we are aware of a region of overlap, matching may always be done, for instance by using Lemma (3.7.2).

This brings us to one of the key problems of perturbation theory, which is how to a *priori* justify the existence of a domain of overlap between two approximations. We shall now talk about some heuristic theories in this regard.

3.8.7 Formal limits of equations and formal domains of validity

We defined the η-limits of functions as well as the permissible domains in §3.8. We shall now discuss related ideas for equations.

We continue to define x_η by $x = \eta/x_\eta$. By including this variable in (3.75), we get

$$\frac{\varepsilon}{\eta^2}\frac{d^2y}{dx_\eta^2} + \frac{1}{\eta}\frac{dy}{dx_\eta} - a - 2\eta b x_\eta = 0 \qquad (3.100)$$

---

[10]In subsequent sections we shall see that matching is used not only to fix the values of certain constants but also to determine the function $\beta_j(\varepsilon)$ of (3.67) and (3.68).

[11]Keep in mind that using intervals with shifting endpoints is essential while discussing the matching theory. the domain within which an approximation is valid. The entire basis for matching would be lacking if we had relied solely on the fact that f_0 is true for any interval $[x_0, 1]$, $x_0 = const. > 0$, Additionally, with the aid of the extended idea of interval, the meaning of such ambiguous sentences as f_0 is valid "away from the origin" and g_0 "near the origin" has now been clarified.

Each term in ε has a *formal order* that is determined by the stated function of ε. For instance, since we consider d^2y/dx_η^2 to be formally of order unity, the first term is of order $\varepsilon\eta^{-2}$. Similarly, x_η, is considered to be of order unity formally; hence the last term is of order η. Since we can multiply (3.100) by any function of e, just the relative order of terms matters. Now that the highest order terms (or term) in the equation (3.100) must have an order of unity, we first multiply this equation by a suitable function of ε and then we may determine the equation's limit as ε approaches zero.

It should be noted that in this limit, x_η is fixed, and the derivatives d^2y/dx_η^2, and dy/dx_η are also formally unaltered. We are actually using equation (3.75a) to apply the limit process $\lim_\eta$. We use the term "formal" since in reality as ε goes to zero, dy/dx_η, a function of both x_η, and ε, will change; whether or not this limit exists is unimportant to us. Wherever functions of ε are explicitly stated, the limit must exist. As a result, we must demand that q and e be compatible, i.e., $1 \cdot \eta^{-1}$ and $\varepsilon \cdot \eta^{-1}$ must tend to definite values, possibly infinity, just as ε tends to zero. We can rewrite the resultant equation in x after the limit has been carried out.

Depending on the o-class of η we then find the following formal limits[12] of (3.75a) (to avoid irrelevant complications we take $\eta = 1$ and $\eta = \varepsilon$ in deriving following equations respectively):

$$\text{ord } \eta > 1: \quad -2bx \qquad\qquad = 0, \qquad (3.101a)$$

$$\text{ord } \eta = \text{ord } 1: \quad dy/dx - a - 2bx \qquad = 0, \qquad (3.101b)$$

$$\text{ord } \varepsilon < \text{ord } \eta < \text{ord } 1: \quad dy/dx \qquad\qquad = 0, \qquad (3.101c)$$

$$\text{ord } \eta = \text{ord } \varepsilon: \quad \varepsilon\frac{d^2y}{dx^2} + \frac{dy}{dx} \qquad = 0, \qquad (3.101d)$$

$$\text{ord } \eta < \text{ord } \varepsilon: \quad d^2y/dx^2 \qquad\qquad = 0 \qquad (3.101e)$$

When η belongs to the function class on the left, the equation on the right is produced.

We observe that the formal η-limit of (3.101b) also yields (3.101c) if η is such that $\lim_\eta$, applied to the "full" equation (3.75a) yields (3.101c).

---

[12]The possible limiting equations are presented below in an oversimplified form. (i) Prior to applying the limits, one may introduce a coordinate change of the form (3.64). An illustration of this is the change brought on by Bush, which is covered ahead in §11 (ii) The dependent variable's magnitude can be thought of in multiple orders. The stream function treatment in flow at high Reynolds numbers is the traditional example (see, e.g., Van Dyke (1964)). The corner-layer, which Kaplun actually presented, in the model equation covered by Cole (1968, p. 32ff.), is an intriguing example.

(3.101b) is "rich enough" to contain (3.101c) in a sense. We define the following definitions.

Definition 3.8.1. *If E is an equation and $\lim_{\eta_1}, E = E_1, \lim_{\eta_2} E = E_2$ and also $\lim_{\eta_2} E_1 = E_2$, we say that E_1 contains E_2 (relative to E).*

Definition 3.8.2. *The formal domain of validity of an equation F, relative to the "full" equation E, is the ord η such that $\lim_{\eta} E$ is either F or an equation contained in F.*

Given the definition provided above, the validity of (3.101b) is

$$\mathfrak{D}_1 = \{\eta | \text{ord } \eta > \text{ord } \varepsilon\} \tag{3.102a}$$

and that of (3.101d)

$$\mathfrak{D}_1 = \{\eta | \text{ord } \eta < \text{ord } 1\} \tag{3.102b}$$

(The notation used in (3.94a) and (3.95a) is the same.) It might be argued that (3.101a) is an absurd equation. But the key here is that the expression

$$\varepsilon \frac{d^2 y}{dx^2} + \frac{dy}{dx} + a - 2bx \equiv \frac{\varepsilon}{\eta^2} \frac{d^2 y}{dx_\eta^2} + \frac{1}{\eta} \frac{dy}{dx_\eta} + a - 2b\eta x_\eta$$

The final term is greater in order than the preceding terms given $1 \ll \eta$.

3.8.8 Outer and inner limits and equations

The two key equations are clearly (3.101b) and (3.101d); one or both of them contain each of the other three equations, but neither of them are contained in the other. We'll refer to them as the principal equations. They are acquired through the aforementioned limit methods. We choose the function $\eta = 1$ to serve as an example of ord 1, and the associated x_η, is then x. Similarly, we consider ε to be a representation of ord ε, and x_η to be x^* according to definition in (3.80). We define

$$outerlimit : \lim_{o} utF(x, \varepsilon) = limF \quad as \quad \varepsilon \downarrow 0, x \text{ fixed}; \tag{3.103a}$$

$$innerlimit : \lim_{o} utF(x, \varepsilon) = limF \quad as \quad \varepsilon \downarrow 0, x^* = x/\varepsilon \text{ fixed}; \tag{3.103b}$$

We refer to the outer and inner equations as (3.101b) and (3.101d), respectively, and the outer and inner variables as x and x^*. Principal limits are the outer and inner limits; *intermediate* limits are the $\lim_{\eta}$, with $\eta \ll \eta \ll 1$. Because they provide the two principal equations in this instance, we can identify the principal limits. Now that we know that (3.78a) and (3.83a) are the primary equations, we can use them to determine f_o and g_o, respectively. Additionally, it is now clear that the preceding assumption that the border layer has a thickness of ε depends on the fact that the inner limit is principal.

There is a hierarchy within the principal boundaries. Since f_0 can be established independently, but g_0 must be obtained by matching with f_0, the outer limit is in a sense the "first" limit. The hierarchy in a physical problem is frequently natural. The solution to the equation represents an unperturbed state. A perturbation of magnitude ε is introduced; this can refer to a change that is physically made in an experiment or that is theoretically taken into account for a small effect that the theory had previously overlooked. By including the expression $\varepsilon(d^2y/dx^2)$ in the equation (3.8.7) for the unperturbed state, the effect of the perturbation is represented. The fact that even if the perturbation is of magnitude ε, its effect could be of order unity, though only in a very small region, is typical for a layer-type singular perturbation problem. Since a new boundary constraint has been imposed there that wasn't imposed on the solution for the state f_0, the effect of the perturbation in our example is of order unity close to the origin as stated by $g_0(x^*)$. Thus, the choice of the first principal equation is frequently made solely on the basis of physical factors.[13] The study of flow at high Reynolds numbers in §3.12 serves as an illustration of the above concepts; in fact, Friedrichs (1953) developed equation (3.75), with $b = 0$, as a model equation for this situation.

Returning to our original problem of heuristic justification for our asymptotic construction, we observe that while the matching principle has been justified for the outer and inner equations, it has not been explained for the matching principle. We've already seen that if we can identify an a priori reason for each of their respective domains of validity, matching of f_0 and g_0 can be explained a priori. Again, we turn to a study of limits of equations.

3.8.9   Heuristic principle for domains of validity

The following heuristic principle attributed to Kaplun will serve as the foundation for our discussion.

Principle 1. *If y is a solution of the equation E and E^* is an approximate equation, then there exists a solution y^* of E^* whose formal domain*

---

[13]Thus the first principal (outer) limit is characterized both by the fact that when applied to the equation it gives a distinguished approximated equation (as described above) and when applied to the solution it gives the unperturbed state. In a reasonable physical theory these two characterizations are equivalent. As will be seen from examples ahead the inner limit need not correspond to a distinguished equation.

of validity covers the actual domain of validity of E^* (as an approximation to E).

We will address significant variations to this principle in the sections that follow. The above formulation is merely sort of a starting point for the conversation. When small terms have significant integrated effects, it is no longer true.[14] The domains of validity of (3.101b) and f_0 on the one hand, and the domains of validity of g_0 and (3.101d) on the other hand, perfectly coincide in our example.

Earlier, we used the comparison to the exact solution to support the asymptotic results. Now that we have adequate information, we can justify something a priori. We'll go over this utilizing the ideas from above. We apply the different limits to equation (3.75a). The two principal limits and the two principal equations (3.101b) and (3.101d) follow naturally from this. We note that the total range $[0,1]$ is covered by all of these equations, although this is not the case for any single equation. Since functions of order unity are included in the domain of equation (3.101b), it is true for any finite x. As long as $\eta \ll 1$, equation (3.101d) is valid on any interval between $[0,\eta]$ for all infinitesimal values of x. For infinitesimal values of x, provided their order of magnitude is greater than ε, equation (3.101b) is likewise formally valid. The overlap domain is

$$\mathfrak{D}_1 \cap \mathfrak{D}_2 = \{\eta | \varepsilon \ll \eta \ll 1\}. \tag{3.104}$$

The heuristic principle (1) is now in use. Therefore, (3.101b) must have a solution in order to be valid in the domain of (3.102a). This has to satisfy the boundary condition at $x = 1$ since, as was previously mentioned, it encompasses the point $x = 1$. This entirely determines the answer to be

$$f_0 = (1 - a - b) + ax + bx^2 \tag{3.105}$$

The boundary condition $x = 0$ must also be satisfied by the solution g_0 to equation (3.101d). This results

$$g_0 = C(1 - e^{-x^*}), \tag{3.106}$$

Since it cannot be assumed a priori that the domain of validity of g_0 includes finite x, the constant cannot be determined by the condition at

---

[14]The basic assumption underlying perturbation methods is a sometimes very vague notion that "neighboring equations have neighboring solutions." The heuristic principle (1) and modifications thereof are examples of this. We can formalize clearly the idea of solutions being close (as was done in §3.8). However, the basic difficulty is to decide when equations are close; this would involve an estimate of integrated effects. These difficulties will be discussed in §3.10 and §3.11

$x = 1$. However, given that we have determined the domain in which f_0 and g_0 overlap, we may use Lemma (3.7.2). As previously demonstrated (see (3.96)), this results in a correct matching.

3.8.10   Higher order terms

The development of higher order concepts may likewise be justified using the heuristic principles discussed above. We write,

$$y = f_0 + \varepsilon \tilde{f}_1. \tag{3.107}$$

Then (3.75a) gives

$$\varepsilon \frac{d\tilde{f}_1}{dx} + \varepsilon \cdot 2b + \varepsilon^2 \frac{d^2 \tilde{f}_1}{dx^2} = 0 \tag{3.108}$$

This is an exact equation. We approximate to $\tilde{f}_1$ by f_1 obeying

$$df_1/dx + 2b = 0 \tag{3.109a}$$

(3.102a) continues to provide the domain of validity of order ε. It follows from (3.75c) that the appropriate boundary condition is

$$f_1(1) = 0, \tag{3.109b}$$

and hence,

$$f_1 = 2b - 2bx. \tag{3.110}$$

Similarly we put

$$y = g_0 + \varepsilon \tilde{g}_1 \tag{3.111}$$

which gives

$$\varepsilon^2 \frac{d\tilde{g}_1}{dx^2} + \varepsilon \frac{d\tilde{g}_1}{dx} - a - 2bx = 0 \tag{3.112a}$$

The exact boundary conditions are

$$\tilde{g}_1(0) = 0, \tag{3.112b}$$
$$\tilde{g}_1(1) = (a + b)/\varepsilon \tag{3.112c}$$

We observe that the problem offered by (3.112) is an exact reformulation of the original problem (3.112), taking into account the definition (3.111) and the expression for g_0 provided by (3.84). We will now attempt to directly address the problem (3.112) using the techniques originally used in relation to the problem (3.112). We use a function g_1 that approximates to $\tilde{g}_1$ and solves a similar problem to (3.101d) to obtain the inner solution:

$$\varepsilon^2 \frac{d^2 g_1}{dx^2} + \varepsilon \frac{dg_1}{dx} - a = 0 \qquad (3.113a)$$

And the boundary condition

$$g_1(0) = 0, \qquad (3.113b)$$

We anticipate that a matching boundary condition will take the place of the second boundary condition. The related outer equation, however, cannot be created using our earlier, more restricted method of limit processes. (3.112a)'s outer limit does not provide a differential equation. In fact, under any $\lim_\eta$, with $\varepsilon \ll \eta$, both terms containing derivatives disappear. On the other hand, $\lim_\eta$ produces the inner equation (3.113a) if $\eta = \varepsilon$. However, an outer approximation of the form is assumed as

$$\tilde{g} \simeq q_0/\varepsilon + q_1, \qquad (3.114)$$

then applications of the outer limit yield

$$\frac{dq_0}{dx} - a - 2bx = 0, \qquad (3.115a)$$

$$q_0(1) = a + b, \qquad (3.115b)$$

$$\frac{d^2 q_0}{dx^2} + \frac{dq_1}{dx} = 0, \qquad (3.115c)$$

$$q_1(1) = 0 \qquad (3.115d)$$

with solution

$$q_0 = ax + bx^2, \qquad (3.116a)$$

$$q_1 = 2b(1 - x). \qquad (3.116b)$$

The answer provided earlier by (3.91) is the one that matches with (3.116a,b) and is the solution g of (3.113a,b). The outer boundary condition (3.112c) for $\tilde{g}_1$ suggests the form (3.114). Developing the equation for q_0 is equivalent to applying the upper limit to (3.112a) keeping $\varepsilon \tilde{g}_1$ fixed. When x is of order unity, the dependent variable is of order ε^{-1} rather than of order unity , hence keeping $\varepsilon \tilde{g}_1$ constant is important.

We observe that by definition, in the same domain where g_1 approximates to $\tilde{g}_1$ to order unity, $g_0 + \varepsilon g_1$ approximates to y to order ε. The actual domain of validity of g_1 is $\mathfrak{D}_3 = \{\eta | \eta \ll \varepsilon^{1/2}\}$, as demonstrated by (3.95b). On the other hand, it appears that (3.113a)'s formal domain of validity is $\mathfrak{D}_2 = \{\eta | \eta \ll 1\}$. It appears that this conflicts with our heuristic principle (1). We are looking for the "effect" of the expression $-2bx$, but we need to take the principle's underlying premises into consideration. This effect in a linear equation is completely independent to the effect of the term $(-a)$.

In order to compare $(-2bx)$ with the derivatives, we disregard the third term. In the case of $\varepsilon \ll \eta$ the first derivative predominates, and we discover that $(-2bx)$ is formally smaller than $\varepsilon(d\tilde{g}_1/dx)$ when $\eta \ll \varepsilon^{1/2}$, that is, exactly in the domain $\mathfrak{D}_3$. The majority of $((-2bx)$'s effect) is primarily

$$\frac{1}{\varepsilon} \int_0^x 2bs\, ds = \frac{bx^2}{\varepsilon}.$$

We observe that this is $o(1)$ in $\mathfrak{D}_3$. As a result, in $\mathfrak{D}_3$, $\tilde{g}_1$ is never large (assuming we ignore the effect of the term $(-a))$).

Of course, one would never, in a practical computation, compute an outer approximation to $\tilde{g}_1$ such as is given by (3.114). Our research only served to demonstrate that, when properly construed, the heuristic principle (1) continued to apply to the domain of validity of g_1. Similar discussions might be had about the functions $\tilde{f}_1$ and f_1. However, there are no issues with that situation.

The reasoning used to support the construction of f_k and g_k for $k > 1$ follows a similar pattern.

We note that equation (3.101c) has a solution given as follows, as additional evidence for our heuristic concept, both positive and negative.

$$y \approx l = 1 - a - b \tag{3.117a}$$

Hence, in the same domain where (3.101c) is true, is uniformly valid to order unity. Similar to (3.101e), it has a solution.

$$y \approx m = (1 - a - b)x/\varepsilon, \tag{3.117b}$$

Hence, in the domain where (3.101e) is valid, is valid to order unity. Both solutions are present in the ones that have already been utilised, thus neither is required in the asymptotic construction. In the meaning of Definition (3.8.1), Equation (3.101c) is contained in (3.101b). Then, we can also consider (3.101c) to be a close approximation of (3.101b).

According to our heuristic principle (1), the solution to equation (3.101c) is (3.116), which approximates the value of f_0 uniformly valid to order unity in the domain where (3.101c) is a formally valid approximation of (3.101b). In the same way, the solution (3.116) is an approximation to g_0 in the domain where (3.101c) approximates (3.101d). In this sense, both f_0 and g_0 contain the functions of (3.116). The same holds true for (3.101e), which is contained in (3.101d) and has a solution in the form of m of (3.117b), which is contained in g_0. The function (3.117b) does not match with the function f_0; it provides the initial slope at $x = 0$, but it does not level off to a constant that can be matched with the constant part of f_0 for small x. Since the domains of validity of (3.101b) and (3.101e) do not intersect, there is no justification for overlap.

3.8.11   Composite expansion

Although neither expansion is uniformly valid throughout the entire interval, we have found two expansions that combined cover the entire interval $[0, 1]$. As a result, we "in principle" have a uniformly valid solution. We may want to construct one expansion that is actually uniformly valid across the entire interval $[0, 1]$ for practical reasons, which is always a top priority in applied mathematics:

$$y \simeq \sum \varepsilon^k h_k(x, \varepsilon). \tag{3.118}$$

Throughout the entire interval $[0, 1]$, h_0 should be uniformly valid to order unity. This will be accomplished if we can create h_0 in such a way that f_0 and g_0 approximate h_0 to order unity in the same domains (or at least in domains encompassing $[0, 1]$) where they approximate the exact solution to order unity. There are various methods for constructing such an h_0. The additive method is the one that works best in this situation. [15] One puts

$$h_0 = f_0 + g_0 - (\text{common part of } f_0 \text{ and } g_0). \tag{3.119}$$

The constant $1 - a - b$ is the common component, which results in

$$h_0 = (1 - a - b)(1 - e^{-x/\varepsilon}) + ax + bx^2. \tag{3.120}$$

The explanation is that we write

$$y = f_0 + (\text{error}) + O(\varepsilon). \tag{3.121}$$

---

[15] For other methods, see Van Dyke (1964).

Only for $x = O(\varepsilon)$ does f_0 have an error of order unity. Consequently, an equation for the error to order unity is required.

$$z = y - f_0 \tag{3.122}$$

which holds true in a certain domain $\eta \ll \eta_0$ where η_0 is a certain function $\ll \varepsilon$. Using the justification provided above, it is found that a first order approximation to z is

$$z \simeq -(1 - a - b)e^{-x/\varepsilon} \tag{3.123}$$

which gives (3.120).

3.8.12   Position of boundary layer

We have implicitly assumed up to this point that the boundary layer starts at $x = 0$. Instead, suppose it happens at point x_0. Then, we define x_η, by $\eta x_\eta = x - x_0$, in particular, x^* by $\varepsilon x^* = x - x_0$. Following the directions above brings us to the boundary-layer solution $Ae^{-(x-x_0)/\varepsilon} + B$. Since ε is positive, the exponential will, nevertheless, go towards infinity when $x < x_0$. If $x_0 > 0$, matching is then impossible. The boundary layer arises at $x_0 = 1$ if $\varepsilon < 0$, as shown by the same argument.

3.8.13   Inner and outer expansions and limit-process expansions

We have constructed two asymptotic expansions of y:

$$y \simeq \sum_{k=0} \varepsilon^k f_k(x). \tag{3.124a}$$

$$y \simeq \sum_{k=0} \varepsilon^k g_k(x^*)., \tag{3.124b}$$

Except for close to the origin, the first expansion which is the *outer expansion* is valid; the second which is the inner expansion is valid close to the origin. These are the only two principal expansions that are required because their domains of validity span the entire range $[0, 1]$. More specifically, for both outer and inner expansions, we need the following: The overlap domain, which connects the domains of validity to order unity of f_0 and g_0, must not be empty, and the union of these domains must span the full range $[0, 1]$. The order ε domains of validity for $f_0 + \varepsilon f_1$ and $g_0 + \varepsilon g_1$ should share the same property. These conditions apply equally to the two expansions. However, as was previously stated, it is f_0 that is perturbed, and g_0 defines the outcome in the thin region where the outcome of the small perturbing component is of order unity. In light of the fact that (3.124a) is the first principal expansion, it follows that f_0 is determined before g_0.

From the exact solution, or from (3.124) we find

$$\lim_{out} y = f_0. \tag{3.125a}$$

$$\lim_{out} \frac{y - f_0}{\varepsilon} = f_1, \tag{3.125b}$$

and by applying the outer limit repeatedly, we are able to achieve $f_0 + \varepsilon f_1 + \dots$. An expansion of this kind is referred to as an outer-limit expansion. Similar to this, applying the inner limit repeatedly results in the inner-limit expansion $g_0(x^*) + \varepsilon g_1(x^*) + \dots$.

The outer-limit expansion in the given example is actually the same as the outer expansion (the same is true for the inner expansion). But they are always conceptually distinct, and the examples in the sections that follow will demonstrate that they might also differ. The inner-limit expansion shouldn't coincide with the outer-limit expansion for any explanation. The method employed here, however, which involves applying limit processes to equations (rather than to functions) and then examining the domains of validity of these equations' solutions, provides heuristic justifications for why the outer and inner expansions constructed from such solutions do match.

3.8.14   Note on terminology

The idea of characterising an expansion by its scope of validity was highlighted in Kaplun and Lagerstrom (1967). While "outer expansion" was still used to refer to a limit process expansion, the phrase "intermediate expansion" was used in the same way as "inner expansion" here. To identify the expansion that corresponds to the outer one but is determined by its scope of validity, a phrase must be established. Furthermore, the term "intermediate expansion" may be used incorrectly to refer to an expansion produced by repeatedly applying a certain intermediate limits. In this context, "inner" and "outer expansions" refer to expansions that are identified by their respective domains of validity. Expansions that include limits are clearly labelled as such. Although this language is different from what is currently used in the literature, the change was thought to be desirable for the reasons listed. Additionally, it highlights the fact that expansions defined by their domains of validity are more fundamentally significant than expansions defined by limit processes.

3.8.15 Techniques of matching

The existence of a domain of overlap serves as the foundation for the matching of two asymptotic expansions, as was previously discussed. In order to really use this, one may write out

$$d_0 = f_0 - g_0, \tag{3.126}$$

where f_0 is determined by (3.79) and g_0 by (3.84a). It is instantly clear that only by adding $C = 1-a-b$ can d_0 tend to zero under an intermediate limit. In the same way,

$$d_1 = (f_0 + \varepsilon f_1) - (g_0 + \varepsilon g_1) \tag{3.127}$$

And it is clear that for the η-limit of $d_1 \varepsilon^{-1}$ to go to zero under some intermediate limit, the constant C_1 in (3.88) must be set to $2b$. We can see that the matching rule (3.89) and the ambiguous reasoning of (3.83c) provided immediately after that rule are both made precise and justified using the technique of the intermediate limit. The connection is that x tends to zero and x^* tends to infinity under an intermediate limit. The x_η, notation should only be used in really difficult instances or for teaching purposes. Additionally, there is no requirement to use a particular η. Practically speaking, one typically adopts an experimental method rather than attempting to predict the degree of overlap in advance. One assumes that there is overlap as long as there is some η-region where the η−limit leads the difference to vanish. However, as will be shown in the sections that follow, there are situations when it is necessary to proceed with a little more caution. Formal matching success does not always imply that the matches were done correctly.

This explanation of the reasons why asymptotic expansions match results in a fairly simple technique for performing the matching.[16]

The earlier method[17] of swapping the outer and inner limits can be related to the technique of applying the intermediate limit to the difference of two partial expansions. For example, (3.83c) might be stated similarly,

$$\lim_{out} \lim_{in} y = \lim_{in} \lim_{out} y \tag{3.128}$$

Additionally, one finds that by applying the inner limit twice (i.e., repeatedly, as shown by (3.125)) to $f_0 + \varepsilon f_1$,

---

[16] A large number of examples are found in Cole (1968). The basic ideas as well as the technique were introduced by Kaplun in the mid-fifties.

[17] Some references are given in Lagerstrom (1957).

$$(1 - a - b) + \varepsilon(ax^* + 2b),$$

whereas applying the outer limit twice to $g_0 + \varepsilon g_1$ (with g_1 given by (3.88)) gives

$$(1 - a - b) + ax + \varepsilon C_1.$$

Requiring above both equation to coincide we find the correct result

$$C_1 = 2b$$

.

We can see that both techniques give the same outcome for the example in the current section. By taking into account a generalisation of the current problem, we shall demonstrate why this is the case.

The following is the general framework of our problem. We have an outer expansion that is also an outer limit expansion

$$y \simeq \sum_{j=0} \varepsilon^j f_j(x). \tag{3.129a}$$

where each f_j has a Taylor expansion

$$f_j(x) = \sum_{k=0}^{\infty} f_{jk} x^k. \tag{3.129b}$$

The terms that make up the "error," or y minus the outer expansion, die away exponentially for large values of x^*. To any order ε", we can ignore these terms by matching by an intermediate limit (where $\varepsilon \ll \eta$) or by repeatedly applying the outer limit to the inner expansion. Therefore, for matching purposes, the inner expansion—which is also an inner-limit expansion—is effectively a rewrite of the outer expansion in inner variables. We must employ the Taylor expansion (3.129b) of the outer expansion regardless of whether we use intermediate limits or rewrite in inner variables. Consequently, we must think about the terms of the form.

$$T_{jk} = f_{jk}\varepsilon^j x^k = f_{jk}\varepsilon^j \eta^k x_\eta^k = f_{jk}\varepsilon^{j+k} x^{*k}. \tag{3.130}$$

Consider now matching to order ε^n. Define

$$A_n = \sum_{j=0}^{n} \varepsilon^j f_j(x) = n + 1 \text{ terms of the outer expansion} \tag{3.131a}$$

$$B_n = n + 1 \text{ terms of the inner expansion,} \tag{3.131b}$$

$$A_n^* = n + 1 \text{ terms of the inner-limit expansion of } A_n, \tag{3.131c}$$

$$B_n^* = n + 1 \text{ terms of the outer-limit expansion of } B_n, \tag{3.131d}$$

We see that T_{jk} will occur in B_n if and only if

$$j + k = m \leqq n \tag{a}$$

It will then be written

$$f_{jk}\varepsilon^m x^{*k}. \tag{b}$$

The same criteria apply for its occurrence in A_n^*. In reality, by ignoring exponentially small terms, we can observe that B_n, is the same as A_n^* and is made up completely of rewriting terms like (b). On the contrary, consider a term like (b) where $m \leqq n$. In outer variables, it is rewritten as

$$f_{jk}\varepsilon^{m-k} x^k. \tag{c}$$

This term appears in B_n^* because $0 \leqq m - k \leqq n$. Thus, we find the matching criterion.

$$A_n^* = B_n^* \tag{3.132}$$

Instead, let's think about matching by an intermediate limit

$$\lim_n \varepsilon^{-n}(A_n - B_n) = 0 \tag{3.133}$$

The two mechanisms make the limit zero.

1. When the difference is formed, some terms in both A_n, and B_n must cancel identically. The terms T_{jk} with $j + k \leqq n$ are plainly those.

2. Even after division by ε^n, other terms tend to zero by themselves. This includes each and every exponentially small terms of B_n.

Additionally, since the nonexponential terms of B_n are obtained by rewriting A_n, the domain of validity of A_n extends arbitrarily close to ord ε. We may see that $\varepsilon^{-n}T_{jk}$ goes to zero under some intermediate limit if and only if $j + k > n$ by taking ord r sufficiently close to ord ε. The most critical case is $j = 0, k = n + 1$, which limits η to be $\ll \varepsilon^{n/n+1}$. Therefore, the region of overlap to order ε^n is

$$\mathfrak{D}_n = \{\eta | \varepsilon \ll \eta \ll \varepsilon^{n/n+1}\}. \tag{3.134}$$

We can see from this that the terms that are still present in the formation of A_n^* and B_n^* are precisely those that must cancel out identically in the difference expression in (3.133). The terms that are eliminated while creating A_n^* and B_n^* are precisely those that in (3.133) tend to zero on their own under an appropriate η-limit.

At the end of §11., we shall talk more about matching techniques.

3.9   Second model equation

[18] As an Example we take a function $y(x; \varepsilon)$ defined by the equation

$$\frac{d^2y}{dx^2} + \frac{n-1}{x}\frac{dy}{dx} + y\frac{dy}{dx} = 0 \tag{3.135a}$$

and the boundary conditions

$$y = 0 \text{ at } x = \varepsilon; \tag{3.135b}$$

$$y = 1 \text{ at } x = \infty. \tag{3.135c}$$

The cases $n = 1, 2, 3$ will be discussed in detail.

We will provide an ambiguous but very useful interpretation for (3.135a). The spherically symmetric function in n dimensions is represented by the Laplacean in the first two terms, where x is the radial variable. If we consider y to be the temperature, then (3.135a) is the equation for the equilibrium temperature distribution, with the term $y(dy/dx)$ denoting a hypothetical nonlinear heat loss. The temperature is equal to one at infinity and zero at the surface of a small n-sphere of radius ε, according to the boundary conditions. According to our interpretation, this is a perturbation problem: The temperature is exactly one everywhere when the sphere is absent. The addition of a cool ($y = 0$) sphere causes this scenario to change. Given that the sphere's radius is $o(1)$, we anticipate that the perturbation will be small (for $n = 2, 3, \ldots$) , with the exception of the region close to the surface where the boundary condition causes the temperature to be zero. The temperature will return to zero everywhere if the sphere vanishes and the radius ε shrinks to zero. To put it more exactly, if one measures the temperature at a constant position (in terms of x), the sphere's impact will disappear as ε goes to

---

[18]The problem discussed here was originally introduced as a model for the asymptotic theory of incompressible flow at low Reynolds numbers as given in Kaplun and Lagerstrom (1957) and Kaplun (1957). Equation (3.135) corresponds to the Navier-Stokes equations (see §12) for viscous flow in n dimensions, (3.140b) to the Stokes equations and (3.145) to the Oseen equations in three dimensions and (A.la) to the Oseen equations in $m - 1$ dimensions. Previously the Stokes equations had been considered as a linearization of the Navier-Stokes equations. Kaplun's analysis of the asymptotic nature of the equations showed that the linearity was a coincidence and would no longer occur for compressible flow (Kaplun (1957)). The asymptotic derivation of the nonlinear Stokes equation for compressible flow is given in Lagerstrom (1964). The model equation of §11 was introduced to illustrate certain mathematical complications occurring in the compressible case. While the "Stokes equation" (3.140b) is linear for the present example, the corresponding equation in §11, namely (3.187b), is nonlinear. Much of the material in §10 and §11 is based on Lagerstrom (1961). For further discussion of the same model equations, see Bush (1971) and the references given there.

zero, or more specifically, as y tends to one. However, this convergence is not uniform close to the sphere. An observer whose x-coordinate varies as $\varepsilon C, C = $ const., is expected to observe a strong influence of the sphere in particular, if $C = 1$, He always observes $y = 0$. We thus have a singular perturbation problem. The heuristic argument suggests the following principal variables:

$$x = \text{ outer variable,} \qquad x^* = x/\varepsilon = \text{ inner variable.} \tag{3.136}$$

The convergence to one as described above can now be described as

$$f_0 = \lim_{out} y = f_0 \equiv 1 \tag{3.137}$$

According to the heuristic interpretation, the convergence to unity occurs uniformly in any interval $[x_0, \infty], x_0 > 0$. It should be noted that the terminology is consistent with the conventions stated in (3.103), the first principal limit is since the outer limit. A small sphere is introduced, and this perturbs the condition $y \equiv f_0$.

We take example $n = 1$ into consideration as a negative conformation of the aforementioned claim. An interval of length 2ε is the one-sphere. The one-sphere can be conceptualised as two planes perpendicular to the x-axis, spaced $\pm\varepsilon$ from $x = 0$ three-dimensional space is used. It is obvious that the "sphere" remains as ε goes towards zero. Space is divided into two separate zones by the planes. Physically, the magnitude of e is unimportant; the shift in coordinate $x \to x - \varepsilon$ takes care of it. The precise solution is, in reality,

$$y(x, \varepsilon) = \tanh\left(\frac{x - \varepsilon}{2}\right) = \frac{1 - e^{-(x-\varepsilon)}}{1 + e^{-(x-\varepsilon)}} \tag{3.138}$$

We can show that, even though $f_0 = 1$ is an approximation that holds for x of large order, it does not hold for x of order unity. For ε small, the function (3.138) does indeed have a regular expansion.

Returning to the examples $n = 2, 3, \ldots$, and so on, we will check to determine if a formal analysis of the limiting equations supports our selection of variables. We introduce the variable x, as in previous example , by x_η, by $\eta x_\eta = x$ and rewrite (3.135a) as

$$\frac{d^2 y}{dx_\eta^2} + \frac{n-1}{x_\eta}\frac{dy}{dx_\eta} + \eta y \frac{dy}{dx} = 0 \tag{3.139}$$

and obtain various approximating equations valid in different domains of the functions $\eta(\varepsilon)$:

$$y\frac{dy}{dx} = 0 \text{ for } \eta \gg 1, \qquad (3.140a)$$

$$\frac{d^2y}{dx^2} + \frac{n-1}{x}\frac{dy}{dx} = 0 \text{ for } \eta \ll 1, \qquad (3.140b)$$

The complete equation (3.135a) is derived for r of order unity. In light of the discussion that follows (3.103) the first principal limit produces a distinctive equation, as it should. The equation for the inner limit (x^* fixed), however, is the same as for any other η-limit with $\eta = o(1)$. The boundary conditions, which were previously described, are where its significance really lies. In contrast to Example I, the inner limit will actually have a smaller impact here, and in Example III, even less. The formal domain of validity of (3.140b) must cover all η in order for $\eta = o(1)$, more specifically $\eta = \varepsilon$. We anticipate that there exists a solution g_0 of (3.140b) valid to order unity in the same domain in accordance with the heuristic principle (1). Such a solution must, in particular, meet the inner boundary requirement (3.135b) and be valid at $x = \varepsilon$. It turns out that using the inner variable to specify such a solution makes it convenient. specified that the inner boundary condition is specified at $x^* = 1$, this is just natural. The e-dependence of the terms in the inner expansion, however, will not always take the simple form: (function of ε) times (function of x^*), as we shall explore in more detail later.

The just-described solution g_0 is established inside an integration constant B (which may be dependent on ε) and is

$$g_0 = B_{in}x^* \text{ for } n = 2, \qquad (3.141a)$$

$$g_0 = B(1 - (x^*)^{2-n}) \text{ for } n > 2, \qquad (3.141b)$$

Matching [19] to the outer solution will yield the constant B. For the outer solution, the function f_0 given by (3.137) appears to be a contender. Given that it satisfies the outer boundary criteria, it is without a doubt uniquely determined to be the solution of the first order equation (3.140a. The regions of validity of (3.140a) and (3.140b), however, do not overlap. But suppose that the η such that ord $\eta = ord$ 1 is also included in the

---

[19]In the corresponding problem in fluid dynamics the fact that (3.141a) cannot satisfy the boundary condition at infinity (3.135c) was known as the Stokes paradox. It was resolved when Kaplun pointed out that the perturbation problem is singular and that the Stokes solution (3.141) is an inner solution which must satisfy a matching condition but not necessarily the boundary condition at infinity. An elementary presentation of Kaplun's argument is given in Lagerstrom and Cole (1955, p. 873ff.).

domain of validity of f_0. Theorem (3.7.3), an extension theorem, can therefore be used to demonstrate that the region $\eta \gg \eta_0$ is part of the domain of validity for some $\eta_0 \ll 1$. Then, because of the overlap in the domains of validity of f_0 and g_0, matching is feasible.

We can't only rely on the equations to prove that f_0 is valid for x of order unity. The whole equation (3.135a) serves as the comparable equation. Since (3.135a) is a second order equation, even though f_0 is a solution to this equation that meets the outer boundary condition, it cannot be calculated with absolute certainty. However, we may utilise the extension theorem instead because our earlier heuristic reasoning shown that f_0 is the outer limit of y, valid on any interval $[x_0, \infty)$ with $x_0 > 0$, which means that its domain of validity encompasses the domain $\{\eta | ord\,\eta \geq ord\,1\}$. Since f_0 must be an outer limit in the given example, the extension theorem in connection with this fact demonstrates that it has a sufficiently broad sphere of validity. However, the validity of the equation that g_0 satisfies is taken into account in order to determine its domain of validity. Although it might be an inner boundary, this is not crucial.

We will explain in parenthesis the need for specific reasons about the above-mentioned area of validity of f_0. The "adjacent" but non-overlapping domains of validity of (3.140a)and (3.140b) cannot automatically be used to infer its enlarged sphere of validity or the existence of an overlap domain. In Example I, the functions $l(x, \varepsilon) = 1 - a - b$ satisfying (3.101c) is valid for ord $\varepsilon < ord, \eta < ord\,1$, and the function $m(x, \varepsilon) = (1-a-b)x/\varepsilon$ satisfying (3.101e) is valid for ord $\eta < ord\,\varepsilon$. However, despite the fact that $l(x, \varepsilon)$ satisfies (3.101d), these two functions do not match. Instead, every function matches a nonconstant solution of (3.101d), namely $g_0(x^*)$ as given by (3.84). It should be noted that it is *formally* impossible to match $m(x, \varepsilon)$, multiplied by an unknown constant, with $l(x, \varepsilon)$. f_0 and g_0 can be formerly matched in Example II. But this is merely a required condition—not a sufficient one. The formal success of matching is not sufficient proof of its correctness, as will be shown by Example III.

Now, we'll work out the specifics for examples $n = 3$ and $n = 2$.

3.9.1   The case $n = 3$

The solution of the inner equation (3.140b) is

$$g_0 = 1 - 1/x^* \tag{3.142}$$

By matching with $f_0 = 1$, the constant B of (3.140b) has been assessed to B in this case. When x^* tends to infinity, that is, under any η-limit such that $\varepsilon \ll \eta$, g_0 obviously tends to f_0. placing it now

$$y = 1 + \beta \tilde{f}_1, \qquad \tilde{f}_1 = f_1 + O(\beta), \tag{3.143}$$

where $\beta(\varepsilon) \ll 1$, we find the exact equation for $\tilde{f}_1$ to be

$$\frac{d^2 \tilde{f}_1}{dx^2} + \frac{2}{x}\frac{d\tilde{f}_1}{dx} + \frac{d\tilde{f}_1}{dx} + \beta \tilde{f}_1 \frac{d\tilde{f}_1}{dx} = 0 \tag{3.144}$$

By applying limit processes to this equation we find equations for f_1:

$$\frac{d^2 f_1}{dx^2} + \frac{2}{x}\frac{df_1}{dx} + \frac{df_1}{dx} = 0, \qquad ord\,\eta = ord\,1, \tag{3.145}$$

$$\frac{df_1}{dx} = 0 \text{ for } \eta \gg 1, \tag{3.146a}$$

$$\frac{d^2 f_1}{dx^2} + \frac{2}{x}\frac{df_1}{dx} = 0 \text{ for } \eta \ll 1, \tag{3.146b}$$

Since (3.146a) and (3.146b) are contained in (3.145), this equation is formally valid everywhere, i.e., in the whole interval $[\varepsilon, \infty)$

The solution of (3.145) satisfying $f_1(\infty) = 0$ have the form

$$f_1 = A_1 E_2(x) \tag{a}$$

where E_2 is defined in the appendix. To determine A_1 and β by matching we apply an intermediate limit to the difference[20]

$$d_0 = (1 + \beta f_1) - g_0 \tag{b}$$

Under the intermediate limit, x tends to zero and x^* tends to infinity, therefore we require the expansion of $1+\beta f_1$ for small x and the expansion of g_0 for large x^*. The expanded form of the function g_0 is already available. We learn from the Appendix

$$1 + A_1 \beta E_2 = 1 + A_1 \beta (1/x + In\, x + O(1)) \tag{c}$$

---

[20]Here we determine the second term of the outer expansion by matching with g_0 alone. The more general rule is that the second terms of both the inner and outer expansions have to be determined by simultaneous matching. See Kaplun and Lagerstrom (1957, p. 590) and Kaplun (1957, p. 598).

Thus, if we put

$$A_1 = -1, \qquad\qquad \beta = \varepsilon \tag{d}$$

then

$$d_0 = -\varepsilon In\, x + O(\varepsilon), \tag{3.147}$$

which tends to zero in the entire x-range (although not uniformly, it does not vanish for $x \sim e^{1/\varepsilon}$). Hence,

$$f_0 + \beta f_1 = 1 - \varepsilon E_2(x) \tag{3.148}$$

In §9, we focused on using matching to get integration constants that are not provided by boundary conditions. The expansions (3.67) and (3.68)'s functions $\beta_j(\varepsilon)$ are also determined by actually matching. This was trivial in §9, but the current section offers a number of examples of determining the β_j that are nontrivial.

3.9.2   Higher order terms

If we now put

$$y = g_0(x^*) + \varepsilon \tilde{g}_1(x^*, \varepsilon), \qquad\qquad \tilde{g}_1 = g_1(x^*) + o(1), \tag{3.149}$$

then

$$\frac{d^2 \tilde{g}_1}{dx^{*2}} + \frac{2}{x^*}\frac{d\tilde{g}_1}{dx^*} + g_0\frac{dg_0}{dx^*} + \varepsilon\left(g_0\frac{d\tilde{g}_1}{dx^*} + \tilde{g}_1\frac{dg_0}{dx^*}\right) + \varepsilon^2 \tilde{g}_1 \frac{d\tilde{g}_1}{dx^*} = 0. \tag{3.150}$$

Neglecting terms of formal order ε and using the value of g_0 given by (3.142), we have

$$\frac{d^2 \tilde{g}_1}{dx^{*2}} + \frac{2}{x^*}\frac{dg_1}{dx^*} = \left(\frac{1}{x^*} - 1\right)\left(\frac{1}{x^*}\right)^2. \tag{3.151}$$

If one introduces any x_η, where $\eta x_\eta = x = \varepsilon x^*$, provided $\eta \ll 1$, the omitted terms retain their formal order ε. Therefore, (3.151) is formally valid in this context.The solutions having the following form that meet the inner boundary constraint $g_1 = 0$ at $x^* = 1$

$$g_1 = -\frac{In\, x^*}{x^*} - In\, x^* + B_1 g_0. \tag{3.152}$$

Hence,

$$g_0 + \varepsilon g_1 = 1 - \frac{\varepsilon}{x} + \varepsilon\left(\frac{-\varepsilon In\, x + \varepsilon In\, \varepsilon}{x} - In\, x + In\, \varepsilon\right) + \varepsilon B_1\left(1 - \frac{\varepsilon}{x}\right) \tag{3.153}$$

This should apply to order ε, as expected. According to the formal domains of validity of the f_1 and g_1 equations, we need $\eta \ll 1$ in order to match by using an intermediary η-limit. Therefore, we must identify the equation for f_1 for small x. We learn from the Appendix

$$f_0 + \varepsilon f_1 = 1 + \varepsilon(-1/x - In\, x - (\gamma - 1) + x/2 + O(x^2)). \qquad (3.154)$$

Dividing by ε, the difference between (3.154) and (3.153), should tend to zero with ε. But this is where our construction fails. Nothing in (3.154) can be matched by the term $\varepsilon\, In\, \varepsilon$ in (3.153). This problem cannot be solved by selecting a constant B_1 (if it is assumed to be independent of ε) or of an η-limit, and the unmatched term cannot be eliminated by continuing the expansions to higher orders.

The formal solution is to revert to lower order words. If we add a term $\zeta(\varepsilon)g_s(x^*)$ to equation (3.153), then g will obey the homogeneous inner equation (3.140b) provided $\varepsilon \ll \zeta$. The apparent option is

$$\zeta g_s(x^*) = -\varepsilon In\, \varepsilon g_0(x^*). \qquad (3.155)$$

Then $\varepsilon\, In\, \varepsilon$ disappears from the expansion

$$y \simeq g_0(x^*) - \varepsilon In\, \varepsilon g_0(x^*) + \varepsilon g_1(x^*). \qquad (3.156)$$

Putting $B_1 = 1 - \gamma$ in order to match the term ε times const., we now find that the difference d_1 between (3.154) and (3.156) is

$$d_1 = \frac{\varepsilon^2 In\, x}{x} - \frac{\varepsilon^2(\gamma - 1)}{x} - \frac{2\varepsilon^2 In\, \varepsilon}{x} + \varepsilon O(x). \qquad (3.157)$$

Matching it now formally possible since

$$\lim_{\eta} \frac{d_1}{\varepsilon} = 0 \text{ for } |\varepsilon In\, \varepsilon| \ll \eta \ll 1. \qquad (3.158)$$

Note that the overlap domain is smaller than anticipated.

While we have managed to overcome our challenges, it is crucial to comprehend why our a priori estimations are insufficient in order to comprehend perturbation theory. Here, we encounter the phenomenon of integrated effects and resulting switchback, which, after some further formal calculations, will be explored in more detail later.

3.9.3 Calculation of f_2

If the outward extension is continued formally as

$$y \simeq 1 + \varepsilon f_1 + \varepsilon^2 f_2 \qquad (3.159)$$

then f_2 obeys (A.3) of the Appendix whose solution is

$$f_2 = F_2 + A_2 E_2, \qquad (3.160a)$$

For small values of x

$$f_2 = \left(-\frac{In\,x}{x} - \frac{\gamma+2}{x} - 3In\,x + O(1)\right) + (2\gamma+1)\left(\frac{1}{x} + In\,x + O(1)\right),$$
$$(3.160b)$$

where the constant A_2 has been evaluated by $2\gamma + 1$.

3.9.4 Switchbacks, integrated effects

Up until it was discovered that (3.153) and (3.154) did not match, nothing in our formal expansion procedure indicated that an intermediate term of order ε, in ε was required. Switchback are terms like (3.155) whose existence is inferred from the inspection of formally higher order terms like (3.152). In singular perturbation constructions, the switchback phenomena is particularly prevalent. [21]

In earlier sections, we saw an example of how a term's impact could be greater than the order's explicit indication. Over a sufficiently enough time span, the term of order ε in (3.64a) produces an effect of order unity. We can write the fundamental equation for the present problem as

$$\frac{d^2y}{dx^{*2}} + \frac{2}{x^*}\frac{dy}{dx^*} = -\varepsilon y\frac{dy}{dx^*} \qquad (3.161)$$

Consider adding the conditions that, for x^* large, y should correspond, to appropriate order, with the outer expansion that is considered to be known in addition to the boundary condition at $x = \infty$ and (instead of the boundary condition at $x = \infty$). Assuming we have the correct solution for $y, y = f(x^*, \varepsilon)$, put its value in the right side while leaving y on the left-hand side as an unknown. This second problem is a linear equation with a well-known forcing term, $-\varepsilon f(df/dx^*)$, and it naturally has the same solution as the first problem. The integrated effect of the right-hand side is the solution of the problem with $\varepsilon \neq 0$ minus the solution of the problem with $\varepsilon = 0$ (which is $g_0(x^*)$). Whether this effect is of order ε or a different order is the crucial question. In our situation, we anticipate that the solution will change by $O(\varepsilon)$ if we use $g_0(x^*)$ in the right-hand side rather than the proper $f(x^*, \varepsilon)$. With a boundary

---

[21]Classical examples from fluid mechanics are the so-called Filon terms discussed by Chang (1961, p. 851ff.) and the $R^2\,In\,R$ term in Proudman and Pearson (1957, p. 255). An artificial but very intuitive example is given in Kaplun (1967, p. 15).

condition at $x^* = 1$ and a matching condition for x^* large, we deal with a problem of the form: a second order linear operator operating on y equals ε. Assume we have a second boundary condition at a location unrelated to ε, say at $x^* = \infty$, and that the relevant Green's function exists. Thus, the formal integral involving the forcing function and the Green's function is the integrated effect of the forcing function. The integrated effect is now of order ε (if we include the degenerate situation of the outcome being identically zero) or the integral does not exist, i.e., the problem has no solution, or its value is ε. (function of x^*). Therefore, the size of integrated effects in any plausible physical problem where we solve a boundary value problem is not a surprise.

Let's go back to the scenario when the outer condition is a matching condition, though. Let us consider our particular case (3.161), where y is changed to $g_0(x^*)$ in the right side. The generic solution that fulfils $y = 0$ at $x^* = 1$ is

$$
\begin{aligned}
y &= C\left(1 - \frac{1}{x^*}\right) - \frac{\varepsilon \, In\, x^*}{x^*} - \varepsilon In\, x^* \\
&= C\left(1 - \frac{\varepsilon}{x}\right) - \frac{\varepsilon^2(\, In\, x - In\, \varepsilon)}{x} - \varepsilon\, In\, x + \varepsilon\, In\, \varepsilon
\end{aligned}
\tag{3.162}
$$

Of course, there is no a priori reason to suppose that C independent of ε. When we match (3.162) with (3.154) to order, we discover that the term In is too large and needs to be eliminated with the help of C. Actually, we had to have

$$
C = 1 - \varepsilon\, In\, \varepsilon + \varepsilon(1 - \gamma)
\tag{3.163}
$$

which of course agree with (3.156).

The consideration of intervals with shifting endpoints in a matching problem, which contain different orders of magnitude of x^* (and x), is crucial for integrated effects. In the corresponding region (where x^* is quite large), the answer would be too huge without the term $-\varepsilon\, In\varepsilon$. However, the addition of this factor increases it beyond what was intended where x^* is of order unity. Since a quantity like dy/dx^* is frequently of physical relevance, this result could be significant.[22]

A technique for assessing integrated effects must be used in a systematic justification of a perturbation approach. Large integrated effects are particularly common when an infinite domain is taken into account or when approximation functions experience singularities like x^{-1}. It is

---

[22] A concrete physical example is the $R^2\, In\, R$ in the drag formula for low Reynolds number flow past a sphere, given in Proudman and Pearson (1967).

obvious that the integrated effects will be in the formal order given in the simple case of example I below.

3.9.5   The case $n = 2$

As Example I we consider the case;

$$g_0 = B(In\, x - In\, \varepsilon) = B\, In\, x^* \qquad (3.164)$$

with $\varepsilon x^* = x$. The difference between g_0 and f_0 is

$$d_0 = B(In\, x_\eta + In\, \eta - In\, \varepsilon) - 1 \qquad (a)$$

If

$$B = -1/In\, \varepsilon + o(-1/In\, \varepsilon) \qquad (3.165)$$

then

$$\lim_\eta d_0 = 0 \text{ for } |In\, \eta| \ll |In\, \varepsilon| \qquad (3.166)$$

In the present example we see that the overlap domain is very small. In particular,

$$\lim_\eta g_0 = 1 - a \text{ for } \eta = \varepsilon^a, \qquad (3.167)$$

in order to prevent the functions $\varepsilon^a, a > 0$, from falling into the overlap domain. However, we observe that the outer limit produces the correct result. Consequently, the overlap domain includes ord 1 and some $\eta \gg 1$ according to the extension theorem. [23]

3.9.6   Higher order terms

as Example II we consider the following case to obtain higher order approximations we proceed as in the case $n = 3$. We assume expansions

$$y = 1 + \alpha \tilde{f}_1, \qquad \tilde{f}_1 = f_1(x) + o(1), \qquad (3.168a)$$
$$y = g_0(x^*) + \alpha \tilde{g}_1, \qquad \tilde{g}_1 = g_1(x^*) + o(1), \qquad (3.168b)$$

where

$$\alpha = -1/(In\, \varepsilon + b), \qquad b = O(1) \qquad (3.168c)$$

we note that

$$e^{-1/\alpha} = \varepsilon e^b. \qquad (3.169)$$

---

[23]This also follows from $|In\, \eta^{-1}| = |In\, \eta|$

Hence the functions $\varepsilon^a, a > 0$, are transcendentally small relative to the α^n. In the exact equation for $\tilde{g}_1$ the highest order terms are

$$\frac{\alpha d^2 \tilde{g}_1}{dx^{*2}} + \frac{\alpha}{x^*} \frac{d\tilde{g}_1}{dx^*} = O\left(\frac{\alpha}{x^{*2}}\right) \tag{3.170}$$

For $x = O(\eta)$ the first term is $O(\alpha\varepsilon^2/\eta^2)$ and the second term is $O(\varepsilon^2/\eta)$. Then, $x = O(\eta)$ is the ratio of the second to the first term. For $\eta \ll \varepsilon^a$, this ratio is transcendentally small when $a > 0$. g_1 will consequently follow the same homogeneous equation as g_0. The same logic will apply to g_2. There is no proof of huge integrated effects because the matching of $g_0(x^*) + \alpha g_1(x^*) + \alpha^2 g_2(x^*)$ and of $1 + \alpha f_1(x) + \alpha^2 f_2(x)$ is formally feasible.

With $m = 1$ and the function f_1 obeying (A.la), the solution is

$$f_1 = -A_1 E(x). \tag{3.171}$$

We write the solution for g_1 as

$$g_1 = B_1 \alpha \, In \, x^*. \tag{3.172}$$

Then, for x small

$$d_1 = 1 - \frac{In \, \varepsilon}{In \, \varepsilon + b} - \frac{\gamma}{In \, \varepsilon + b} + \frac{B_1 \, In \, \varepsilon}{(In \, \varepsilon + b)^2} - B_1 \frac{In \, x}{(In \, \varepsilon + b)^2} + O(x)\alpha$$

$$= \frac{b}{In \, \varepsilon} - \frac{\gamma}{In \, \varepsilon} + \frac{B_1}{In \, \varepsilon} - \frac{B_1 \, In \, x}{(In \, \varepsilon)^2} + O(x)\alpha. \tag{a}$$

The obvious choice is to put

$$B_1 = 0 \qquad \text{and } b = \gamma, \tag{3.173a}$$

i.e.,

$$\alpha = -1/(In \, \varepsilon + \gamma) \tag{3.173b}$$

Then

$$d_1 = O(x)\alpha \tag{b}$$

and

$$\lim_{\eta} d_1/\alpha = 0 \text{ for } \eta = o(1). \tag{3.174}$$

If instead we put $b = 0$, then we must have $B_1 = \gamma$ and the term $B_1 \, In \, x/(In \, \varepsilon)^2$, divided by α, goes to zero only in the very restricted domain where $|In \, \eta| \ll |In \, \varepsilon|$.

The term f_2 obeys (A.3) of the appendix; hence,

$$\begin{aligned}
f_2 &= 2E(2x) - (e^{-x} \cdot E(x)) + A_2 E(x) \\
&= -(1 + A_2)In\,x - 2\,In\,2 - \gamma - A_2\gamma - x\,Inx + O(x).
\end{aligned} \qquad (3.175)$$

This solution has to be matched with a solution valid near the body

$$\alpha^2 g_2 = B_2 \alpha^3 \, In\,x^*. \qquad (3.176)$$

The difference

$$d_2 = (f_0 + \alpha f_1 + \alpha^2 f_2) - (g_0 + \alpha g_1 + \alpha^2 g_2) \qquad (3.177)$$

becomes, with

$$A_2 = -1 \qquad \text{and} \qquad B_2 = -2\,In\,2, \qquad (3.178)$$

$$d_2 = d_1 + \alpha^2[-2\,In\,2 - x\,In\,x + 2\,In\,2 \cdot \alpha(In\,x - In\,\varepsilon) + O(x)]. \qquad \text{(c)}$$

Since $d_1 = O(x) \cdot \alpha$

$$\lim_{\eta} d_1/\alpha^2 = 0 \qquad \text{for} \qquad \eta \ll \alpha. \qquad \text{(d)}$$

Thus,

$$\lim_{\eta} d_2/\alpha^2 = 0 \qquad \text{for} \qquad |In\,\eta| \ll |In\,\varepsilon| \qquad \text{and} \qquad \eta \ll \alpha. \qquad (3.179)$$

The peculiar transformation of the logarithmic function into a constant—for example, $In\,x^*/$-$In\,\varepsilon$ tends to unity for $x = O(\eta)$, where η is in the overlap domain—is a crucial aspect of the discussion of case $n = 2$. An important aspect of the study of layertype problems is that we take into account areas of x of different orders, and that the order of a function of x relies on the order of x, as we saw earlier in the discussion of integrated effects. In x^* is $O(\alpha)$ for $x = O(\varepsilon)$ but $O(1)$ for $x = O(\varepsilon^a), 0 < a < 1$.

According to (3.167), if x declines as ε^a, g_0 generally tends to the constant value $1 - a$. In a sense, all the constants in the range $[0, 1]$ are generated by the logarithmic function g_0. While $\lim_{\eta} y$ for $\eta \ll 1$ produces a continuum of limits, Kaplun observed (in the corresponding fluid dynamics problem) that the same limit processes applied to the equation only produce one equation, namely (3.140b) with $n = 2$. Then, his plan was to find a solution to this equation that had all of these limits but didn't have to be a limit. In fact, the answer is $g_0(x^*)$.

What sense the inner solution perceives the y-value ("temperature") at infinity has just been discussed. Since $y < 1$ for any finite x, it should actually be matched with a somewhat lower value. The correction to order of the temperature at the apparent infinity is provided by the constant term in the expansion of f_1. With the use of a unique definitional trick α, this was resolved. Similarly $\alpha^2 f_2$ gives a correction to order α^2 of the temperature at the apparent infinity the inner term $\alpha^2 g_2$ has to agree with this value in the same sense in which g_0 agreed with the value one. In the inner region, the term $y(dy/dx)$ will be transcendentally small in the sense that it will never appear in the equations for the g_j. However, by the matching conditions, it will indirectly affect the g_j.

We'll now offer a number of comments regarding the outcomes for $n = 2$ and $n = 3$.

3.9.7 Outer and inner expansions versus limit-process expansions

We can use limit processes on these expansions rather than y because partial sums of the inner and outer expansions together are uniformly valid to an order depending on the number of terms considered. The outer and inner limits of y for $n = 3$ are found to be f_0 and g_0, respectively. Repeatedly applying the inner limit, we find

$$\lim_{in} \frac{y - g_0}{\zeta} = g_s(x^*) = g_0(x^*) \tag{3.180}$$

with $\zeta = -\varepsilon \, In \, \varepsilon$. Since

$$\lim_{out} \frac{y - f_0}{\zeta} = 0, \tag{3.181}$$

We could set $f_s = 0$. Then $g_0 + \zeta g_s$ and $f_0 + \zeta f_s$, respectively, are the first two terms of the outer-limit expansion and the inner-limit expansion. As we saw earlier in this section, considering f_0 as an outer limit helped to define its domain of validity. A limit-process expansion does not, however, typically provide any clues regarding the domain of validity. In fact,

$$(f_0 + \zeta f_s) - (g_0 + \zeta g_s) = 1 - (1 - \varepsilon/x) + \varepsilon \, In \, \varepsilon (1 - \varepsilon/x) \tag{3.182}$$

This difference , divided by $\zeta = -\varepsilon \, In \, \varepsilon$, does not tend to zero.

The leading terms of the outer and inner expansions, respectively, are the functions f_0 and g_0 because they provide uniform validity to order unity, i.e., the union of their domains of validity is the complete interval $\varepsilon \leqq x < \infty$. If one is interested in computing $(dy/dx^*)_{x^*=1}$), for example, the term $\zeta g_s(x^*)$ gives higher order validity near $x^* = 1$ and provides a correction to g_0. The terms ζg_s and ζf_s do not, however, give uniform validity to an order higher than unity, as shown by (3.182), meaning that adding these terms does not improve the uniform validity. In fact, εf_1 is the second term of the outer expansion. We disregard f_s. It is identically zero; the only time one needs it for bookkeeping is if they insist on limit-process expansions. Take $\varepsilon g_a(x, \varepsilon) = -\varepsilon \ln \varepsilon g_0(x^*) + \varepsilon g_1(x^*)$ as the inner expansion's second term. The inner and outer expansions together are uniformly valid to order ε if the above term is added to g_0 and εf_1 to f_0. As a result, the inner expansion's second term does not take the form $\beta(\varepsilon) g(x^*)$. The inner expansion doesn't need to take on such a unique form, though, if we free ourselves from the idea that it must be a limit process expansion.

As usual, we find for $n = 2$ that outer limit of y equals one and

$$\lim_{in} y = 0, \qquad \lim_{in}(y - 0)/\alpha = \ln x^*. \qquad (3.183)$$

The inner-limit expansion's first term is therefore zero. We discover that the second term of the inner-limit expansion agrees with the first term of the inner expansion by repeatedly applying the inner limit. Thus, it is clear that the inner and outer limits do not coincide. A priori, we can only predict the following for the domains of validity of limits: Let f be the η_0-limit of a function $z(x, \varepsilon)$. By definition, $\lim_{\eta_0}(z - f)$ is zero, and the functions η_0 must be included in the domain of validity of f as an approximation to z. The extension theorem assures us that η_1 and η_2, $\eta_1 \ll \eta_0 \ll \eta_2$, must exists such that

$$\mathfrak{D}' = \{\eta | \eta_1 \ll \eta \ll \eta_2\} = (ord\, \eta_1, ord\, \eta_2), \qquad (3.184a)$$

is covered by the domain of validity Let $\zeta_0 \ll \eta_0$ and the ζ_0-limit of z be g. When this happens, g is valid in some open order-interval that contains,

$$\mathfrak{D}'' = (ord\, \zeta_1, ord\, \zeta_2) \qquad (3.184b)$$

For a few functions, $\zeta_1 \ll \zeta_0 \ll \zeta_2$ applies. However, there is no requirement that $\mathfrak{D}'$ and $\mathfrak{D}''$ overlap in the extension theorem. The form (3.184a) with $\eta = 1$ had a domain of validity due to the fact that f_0 was an outer limit. This method, however, only worked once, and g_0 had to be determined as a solution whose formal domain of validity overlapped that of f_0 rather than as a limit

The same terms can be seen in both the inner and inner-limit expansions, at least to the order computed here. However, we note that the inner-limit expansion's first term is 0 and its second term is $\alpha \ln x^*$. But $\alpha \ln x^*$ is the first term in the inner expansion. This term, along with the first term in the outer expression, $f_0 \equiv 1$, provides uniform validity to order unity.

3.9.8 Inner limit of the outer expansion

We would anticipate that a solution of the first equation would contain g_0 because the equation for f_1 contains the equation for g_0. Actually, the inner limit of $1 + \varepsilon f_1$ (which follows the same rules) for $n = 3$ equation as f_1 is g_0, and for $n = 2$, the result of applying the inner limit once again is true

More generally, the inner-limit expansion and the inner expansion are produced by regrouping of terms the inner limit to the outer expansion. For $n = 3$, for instance

$$f_0 + \varepsilon f_1 + \varepsilon^2 f_2 + 2\varepsilon^2 \ln \varepsilon E_2 = 1$$
$$+ \left(-\frac{1}{x^*} - \varepsilon \ln x^* - \varepsilon \ln \varepsilon + \varepsilon(1 - \gamma) + O(\varepsilon^2) \right)$$
$$+ \left(-\frac{\varepsilon \ln x^*}{x^*} - \frac{\varepsilon \ln \varepsilon}{x^*} + \frac{\varepsilon(\gamma - 1)}{x^*} + O(\varepsilon^2) \right)$$
$$+ \left(\frac{2\varepsilon \ln \varepsilon}{x^*} + O(\varepsilon^2 \ln \varepsilon) \right) \qquad (3.185)$$

Thus the inner expansion's leading terms can be calculated. (Note that in order to match the switchback term of order ord ($\varepsilon \ln \varepsilon$) of the inner expansion, a switchback term of order $\varepsilon^2 \ln \varepsilon$ has been added to the outside expansion.) This suggests that each partial outer expansion, which is verifiable, satisfies the boundary condition at $x = \varepsilon$ to a certain order. Then, we could wonder why anyone would ever use the inner expansion. One reason is that the inner expansion, so to speak, is one step ahead of the outer expansion if one is just concerned with computing conditions around the sphere. In our physical comparison, a typical quantity to compute may be, for example, $\varepsilon(dy/dx)_{x=\varepsilon}$, which could be the heat transfer at the sphere's surface. Take case $n = 2$. g_0, which is obtained by solving a relatively simple equation and then matching with $f_0 = 1$, can be used to compute the quantity mentioned to the lowest order. The calculation is enhanced by computing f_1 . (In our computations, we discovered an improved value of e, allowing us to select $g_1 = 0$). In order to determine g_2, we need to know f_2. Examining the math reveals that we just require the value of f_2 at zero minus its singular part. This number corresponds to the previously discussed α^2 correction to the temperature at the apparent infinity. The exact same method was used by Kaplun (1957) to determine an improved value of the drag on a circular cylinder in a highly viscous flow. Finding f_2 as a function of x in this situation would have taken substantial numerical computation, however the order unity term of f_2 in the expansion at $x = 0$ may be discovered by evaluating an integral.

It's a coincidence that the inner expansion can be found in the outer expansion. If we change Example II to Example III, which will now be studied, it will no longer be true.

3.10   Third model equation

As Example III we take a function defined by

$$\frac{d^2y}{dx^2} + \frac{n-1}{x}\frac{dy}{dx} + \left(\frac{dy}{dx}\right)^2 + y\frac{dy}{dx} = 0, \tag{3.186a}$$

$$y = 0 \qquad \text{at} \qquad x = \varepsilon, \tag{3.186b}$$

$$y = 1 \qquad \text{at} \qquad x = \infty, \tag{3.186c}$$

The only thing that sets this example apart from Example II is the addition of the term $(dy/dx)^2$ to the equation's left-hand side.

These are the approximate equations.

$$y\frac{dy}{dx} = 0 \qquad \text{for } \eta \gg 1, \tag{3.187a}$$

$$\frac{d^2y}{dx^2} + \frac{n-1}{x}\frac{dy}{dx} + \left(\frac{dy}{dx}\right)^2 = 0 \text{ for } \eta \ll 1, \tag{3.187b}$$

and the full equation for ord $\eta = ord\,1$.

The solution of (3.187b) satisfying the inner boundary condition is

$$n = 2: \qquad g_0 = In(1 + B\,In\,x^*), \tag{3.188a}$$

$$n = 3: \qquad g_0 = In\frac{x^* - B}{x^*(1 - B)} \tag{3.188b}$$

The constant of integration B is no longer multiplicative as it was in the comparable answers for Example II as the inner equation (3.187b) is nonlinear. The two situations are very different from one another in this way.

As in §10, we demonstrate that $f_0 \equiv 1$ is the leading term of the outer expansion. The equation for f_1 is (3.145) and is the same as in §10. Compared to equation (3.186)), the equation for f_2 contains an additional forcing term (caused by the phrase $(dy/dx)^2$ in (3.186)), but it can still be solved with the same ease.

Even if the two cases share a lot of similarities, they also have some key distinctions: The "Stokes equation" (3.187b)is no longer linear, and the "Oseen equation" does not contain it [24]. The inner expansion can no longer be obtained from the outer expansion, as was done in (3.185), as a result of the second difference. We'll discuss the first difference's significant consequences below.

Thus, the leading terms of the outer expansion are

$$f_0 = 1 \qquad \text{for all positive integers}$$

$$f_1 = -AE(x) \text{ for } n = 2, \tag{3.189a}$$
$$f_1 = -AE_2(x) \text{ for } n = 3, \tag{3.189b}$$

The expansion of g_0 for large x^* is required for matching. Since In In x^* would otherwise be the dominant factor in the situation $n = 2$ we can see that B In x^* cannot be huge. It is necessary for B In x^* to tend to $e - 1$ in order for g_0 to tend to one. This results

$$n = 2 : \qquad B = \varphi(\varepsilon)(e - 1), \quad \varphi(\varepsilon) = -1/In\,\varepsilon \tag{3.190a}$$

The domain of overlap is the same as in Example II.
Similarly

$$n = 3 : \qquad B = (e - 1)/e. \tag{3.190b}$$

For $n = 3$, g_0 thus happens to satisfy the condition at infinity; for $n = 2$, it does not.

The second term of the inner-limit expansion in Example II for $n = 2$ matched with f_0. This is no longer true for the present example, which is a significant effect of (3.187b)'s nonlinearity. $g_0(x^*)$'s inner-limit expansion for $n = 2$ begins.

$$g_0(x^*) \simeq 0 + \varphi(\varepsilon)(e - 1)In\,x^* + \ldots . \tag{3.191}$$

---

[24]The same mathematical differences exist between the problems of flow at low Reynolds numbers in the compressible case and the incompressible case respectively. (See footnote 16.)

We can infer that the explicitly stated terms also serve as the start of y's inner-limit expansion. We now see that the *inner-limit expansion and the outer-limit expansions* do not match in the current situation; to be more specific, no partial sum of the inner-limit expansion of y will match with f_0. The absence of a third differentiated limit denoting a buffer layer is not the cause of the matching's failure. The only distinguishable bounds are the inner and outer limits. For $n = 2$, it makes sense a priori that y's inner limit is zero. Consequently, one can predict that the inner-limit expansion will take the form

$$y \simeq 0 + \varphi(\varepsilon)l(x^*) + \dots. \tag{3.192}$$

where $\varphi(\varepsilon) = o(1)$ but is as yet undetermined. The equation for $l(x^*)$ is then

$$\frac{d^2 l}{dx^{*2}} + \frac{1}{x^*}\frac{dl}{dx^*} = 0 \tag{3.193}$$

This equation, in contrast to (3.187b), is invalid in a domain that overlaps the outer limit. It is valid in the area where $y = o(1)$, which is a separate area from the area where $y \sim 1$ is valid.

Take note that formally,$y = 1$ and a solution to equation (3.193) correspond.

$$\varphi(\varepsilon)l(x^*) = \frac{-1}{In\,\varepsilon} In\,x^*. \tag{3.194}$$

Since the element $e - 1$ is missing, this is obviously the incorrect outcome.[25]This proves that formal matching success is not a guarantee of the accuracy of the outcome.

3.10.1   Change of coordinates for $n = 2$

In an interesting modification of the coordinates for $n = 2$, Bush (1971) has made the inner expansion into an inner-limit expansion. He takes

$$t = In\,(1/x). \tag{3.195}$$

Equation (3.186) then become ($n = 2$)

$$\frac{d^2 Y}{dt^2} + \left(\frac{dY}{dt}\right)^2 - exp(-t)Y\frac{dY}{dt} = 0, \tag{3.196a}$$

$$Y = \begin{cases} 0 & \text{at} \quad t = In(1/\varepsilon), \\ 1 & \text{at} \quad t = -\infty \end{cases} \tag{3.196b}$$

---

[25]It has been proved rigorously in Cohen and Lagerstrom (to be published) that g_0 as given by (3.188a) is indeed correct.

where $y(x, \varepsilon) = Y(t, \varepsilon)$.

We will limit our talk because the transformation is significant just for inner expansion. When the inner layer is present, Bush provides the inner variable.

$$t^* = \varphi(\varepsilon)t. \tag{3.197}$$

When looking at the current coordinates, it is especially simple to understand that the final component in (3.196a) will be exponentially small and consequently transcendentally small in the inner region (3.169)). We put

$$g_j(x^*, \varepsilon) = G_j(t^*) \tag{3.198}$$

(As will be clear, the G_j will not explicitly depend on ε). hen the inner expansion is

$$Y \simeq G_0 + \varphi G_1 + \ldots. \tag{3.199}$$

one find

$$G_0 = In[1 + (e - 1)(1 - t^*)], \tag{3.200a}$$
$$G_1 = \gamma(e - 1)(1 - t^*)[1 + (e - 1)(1 - t^*)]^{-1}. \tag{3.200b}$$

(The calculation of G requires matching with the third outer term, $f_2(x)$, which is computed in the same manner as the corresponding f_2 in Example II.)

As a result, G_0 and G_1 do not explicitly contain ε, and each term in the inner expansion now corresponds to a term in the inner-limit expansion. None of the matching heuristics are altered by this knowledge. It does, however, have some benefits for the expansion's actual construction. The steps taken by Bush are as follows. As was previously demonstrated, one first finds go. By using the g_0 form as a guide, one discovers the variables t and t^* such that the explicit reliance of ε in $G_0(t^*)$ will vanish. Thus, it is anticipated that the equation for G_1 will be somewhat less complex than that for g_1. In this instance, we have

$$\frac{d^2 G_1}{dt^{*2}} + 2\frac{dG_0}{dt^*}\frac{dG_1}{dt^*} = 0 \tag{3.201}$$

We can see from (3.201) that g_1 obeys the equation by defining it as the function of x and ε that corresponds to $G_1(t^*)$.

$$\frac{d^2 g_1}{dx^{*2}} + \frac{1}{x^*}\frac{dg_1}{dx^*} + 2\frac{dg_1}{dx^*}\frac{dg_0}{dx^*} = 0 \tag{3.202a}$$

with

$$\frac{dg_0}{dx^*} = \frac{\varphi(\varepsilon)(e-1)}{[1+(e-1)\varphi(\varepsilon)In\,x^*]x^*}. \tag{3.202b}$$

The solution to the equation (3.201), which satisfies the necessary matching and boundary conditions, is

$$G_1 = \frac{\gamma(e-1)(1-t^*)}{1+(e-1)(1-t^*)}. \tag{3.203}$$

When this is rewritten in x^*, the answer to equation (3.202) is

$$g_1 = \frac{\gamma(e-1)\varphi\,In\,x^*}{1+(e-1)\varphi\,In\,x^*}. \tag{3.204}$$

Using the outer variable, one discovers that for large values of x^*,

$$g_1 = \gamma\frac{e-1}{e} + \varphi\gamma\frac{e-1}{e^2}In\,x + O(\varphi^2). \tag{3.205}$$

From the perspective of the current study, Bush's transformation is more interesting for the following two reasons than for the calculations' simplification: Firstly, one need not divide x by a function of ε to create the inner variable; one may instead divide a function of x by a function of ε. One may even employ a more general coordinate change, as stated by (3.64), in place of a function of x. While it is a nonconstructive generalization, this assertion is quite evident. Bush's approach is intriguing since it systematically identifies a particular coordinate change that makes the analytical process simpler, the question of whether this can be done consistently for a wide range of problems is raised. Second, the coordinate shift raises the following problem, which is especially important in the context of the current study: Under the condition that the explicit formula for the derivative of g_0 is not used, (3.202a) can be produced by applying limit processes to the appropriate equations, as illustrated in detail in §10. The third term will, however, vanish in the limit if the explicit expression (3.202b) is employed, and the resulting equation is

$$\frac{d^2\tilde{g}_1}{dx^{*2}} + \frac{1}{x^*}\frac{d\tilde{g}_1}{dx^*} = 0. \tag{3.206}$$

The solution[26] is

$$\tilde{g}_1 = \varphi\gamma\frac{e-1}{e}In\,x^* = \gamma\frac{e-1}{e} + \varphi\gamma\frac{e-1}{e}In\,x. \tag{3.207}$$

---

[26]Both g_1 and $\tilde{g}_1$ match with the "temperature" correction at the apparent infinity as given by the constant term $\gamma(e-1)/e$ of f_2. The discussion after (3.179) applies equally well to the present case.

The two outcomes differ considerably from one another. Similar to Example II, $(dy/dx^*)_{x^*=1}$ may be crucial in a physical problem. While $\varphi^2\gamma(e-1)/e$, φg_1 contributes $\varphi^2\gamma(e-1)$ to this quantity. Higher order terms cannot therefore be used to fix the discrepancy between the two values. However, $\tilde{g}_1$ is the right answer; the justification for this will now be discussed.

3.10.2 Integrated effects

We express (3.202a) as follows to better understand the situation at hand.

$$\frac{d}{dx^*}\left[In\left(x^*\frac{dg_1}{dx^*}\right)\right] = -2\frac{dg_0}{dx^*}. \tag{3.208}$$

We discover that the right side is of order φ by inserting (3.202b). Integrating, however, changes it into a term of order unity. As a result, the integrated impact is of order unity, and the function g_1, rather than $\tilde{g}_1$, provides the right answer.

In §10, immediately following (3.161), we talked about the combined effects of a term of the form (function of ε). (function of x). We discovered that if a matching requirement rather than a boundary condition were imposed, the effect of such a term might be greater than anticipated. There was no question in that situation as to which equation should be used for g_1 (or rather, for g_0). The only question was deciding which answer to employ because neither the one predicted by a limit-process expansion nor by a superficial order estimate was the right one. The main question in the current situation is which equation to use. We actually need to revise how we think of an equation's formal limit. Since (3.167) states that η-limit of φ In x^* is $1-a$ for $\eta = \varepsilon^a$, applying formal limits to (3.208) would result in an infinite number of equations. Determining that φ In x^* is formally of order unity requires us to extend the explanation that follows (3.179) from solutions to equations to all limits. We also observe that g_1 and g_0 are of the same order in the current example.

The heuristic principle is a very helpful starting point for developing solutions, but it needs to be adjusted by taking integrated effects into account, as shown in Examples II and III. In addition, we have learned that our initial assumptions about the limits of equations need to be revised. It goes without saying that it is a tautology to say that integrated effects should be properly taken into account. The switchback effect, which shown in Example II for $n = 3$, is the one that is the simplest to identify. The impossibility of matching told us that the expected form of the expansion was incorrect. The challenges that Example III demonstrates are more subtle. It demonstrates that the formal success of matching does not imply that the solutions truly match. The faulty formula for g_0 provided by (3.194) served as the first example of this. Though, in this case, our heuristic approach produced the desired outcome, and there are no significant integrated consequences. There are more obstacles to discovering g_1. Assume that the matching and boundary conditions are as stated above, and that equation (3.202a) was provided as the original equation with dg_0/dx^* explicitly written . How would one know that g_1 rather than $\tilde{g}_1$ is the right answer if they were trying to establish a first approximation for the value of ε small? The only possible response is that, once the question is asked, the proper answer can be discovered. However, if one is blindly oblivious to terms of order φ and is not alerted of the possibility of large integrated effects, the incorrect answer will be obtained.

3.10.3　Matching reconsidered

The previous method of switching between limit expansions, which was covered at the end of §9, may be justified in some simple cases where it was demonstrated to be almost equal to using the intermediate limit. In simple cases, one can even create a specific rule, such as (3.132). As a broad matching concept, Van Dyke (1964, p. 90) suggested generalising this rule. In a thorough investigation, Fraenkel (1969) demonstrated the circumstances in which this rule is true and also provided several counter examples

Despite the fact that the purpose of the current study is very different from Fraenkel's, certain remarks will be made. *Any* general matching principle that relies on the exchange of inner and outer limit processes has the fundamental flaw of assuming that the inner and outside expansions are genuinely produced by limit processes. In addition to being more fundamental for comprehending matching, the idea of characterizing expansions according to their area of validity is frequently crucial in creating these expansions. In §11, a clear illustration of this is shown. Example II for $n = 3$ is less obvious but equally significant. The second term of the inner expansion is actually the function $-\varepsilon In\,\varepsilon g_0(x^*) + \varepsilon g_1(x^*)$, as stated following (3.182). Of course, one may try to develop rules for applying the inner-limit expansion to the outer expansion (in the sense used here) and comparing it to the inner-limit expansion's outer-limit expansion. Any such rule must define the method for counting terms in addition to the number of terms to be used in the inner and outer expansions.

Another crucial point that has already been raised is that matching is always dependent on the overlap, regardless of the approaches employed. Comparing two approximations that do not share a valid domain is clearly not a good idea. Additionally, the method always involves comparing the inner expansion for large x^* with the outer expansion for small x (as long as x and x^* are the two primary variables, of course). By using various instances, the present study demonstrates how this may be accomplished relatively easily by taking into account the terms' order in the difference between a partial outer and a partial inner expansion.

3.11 Examples from fluid dynamics

Using comparatively simple model equations, some related problems of singular perturbations were treated in earlier parts. These equations can be considered independently of fluid dynamics, despite the fact that they were initially introduced to demonstrate the usage of specific concepts and methods of singular perturbations employed in mathematical fluid dynamics. Using examples directly drawn from fluid dynamics, we'll demonstrate how various specific difficulties can develop under singular perturbations in this section. It often appears impossible to create simple model equations without losing the crux of the problem. Van Dyke (1964) provides other examples.

3.11.1 The Navier-Stokes equations

In this section, the two-dimensional stationary incompressible viscous flow will receive most of our attention (see Lagerstrom, 1964).

$$v = \text{ kinematic viscosity of the fluid,} \tag{3.209a}$$

$$u_0 = \text{ reference velocity,} \tag{3.209b}$$

$$p_0 = \text{ reference pressure,} \tag{3.209c}$$

one may form the reference length of the fluid

$$\lambda = v/u_0. \tag{3.210}$$

We'll use the nondimensionalized by λ, rectangular length coordinate $x = (x_1, x_2)$ to represent it. The velocity u_0 nondimensionalized will be represented by the symbol $q = (q_1, q_2)$. The nondimensionalized pressure, p_0, will be indicated by the letter p.

The nondimensional form of the Navier-Stokes equations is then

$$\sum_{i=1}^{2} q_{i,i} = 0 \tag{3.211a}$$

$$\sum_{i=1}^{2} (q_i q_{j,i} + p_{,j} - q_{j,ii}) = 0 \qquad j = 1, 2 \tag{3.211b}$$

Here $f_{,i} = \partial f / \partial x_i$.

As boundary condition we require that

$$q = 0 \text{ on solid surface.} \tag{3.212}$$

For flow of an infinite medium past a solid we require

$$q = 1, \quad p = 1 \text{ at infinity.} \tag{3.213}$$

Naturally, this indicates that in (3.209b) and (3.209c), we utilised the value at infinity as the reference value.

A solid surface is referred to in the boundary condition (3.212). The Reynolds number, represented here by the letter R, is a nondimensional parameter that can be formed if this surface has the characteristic length L.

$$R = L/\lambda. \tag{3.214}$$

As was covered earlier, if two lengths in a problem are of a different order of magnitude, perturbation techniques may be used, but we must be prepared for the possibility of singular issues.

3.11.2   Flow at low Reynolds numbers

Consider the viscous flow past an object whose usual dimension L is substantially lower than the viscous length λ, i.e.

$$L \ll \lambda \text{ or } R \ll 1. \tag{3.215}$$

As inner coordinate we use

$$x^* = x_d im/L = x/R \tag{3.216}$$

Kaplun is responsible for the fundamental concepts of how to build an asymptotic expansion for small R. We won't go into detail about them here (see Van Dyke (1964) and the references provided there). Our second model equation illustrates the key concepts involved. (Footnote 17 discusses how the second and third model equations relate to fluid dynamics.)

3.11.3   Flow at high Reynolds numbers

Written in x_j^*-coordinates, (3.211) becomes

$$\sum \frac{\partial q_i}{\partial x_i^*} = 0 \tag{3.217a}$$

$$\sum_i q_i \frac{\partial q_j}{\partial x_i^*} + p_{,j} = \frac{1}{R} \sum_i \frac{\partial^2 q_j}{\partial x_i^{*2}}. \tag{3.217b}$$

We set the right side of (3.217b) equal to zero for high values of R. The equations that emerge from this are known as the Euler equations. However, we must weaken (3.212) to because of the equations' lower order.

$$\text{normal component of } \mathbf{q} = 0 \text{ on solid surface} \tag{3.218}$$

It is impossible to calculate the so-called skin friction, a viscous force that the fluid exerts on the body, from a solution of (3.218). Prandtl (1905) noted that from its true value ($= 0$) to the value provided by the solution to equation (3.218), the tangential velocity changes fairly quickly. This transition occurs in a layer of thickness ε known as the boundary layer, where

$$\varepsilon = 1/\sqrt{R} \tag{3.219}$$

Prandtl substituted the x_i^* with the coordinates ξ and η, where $\eta = 0$ is located on the body's surface, in order to examine this layer. If η then is changed to

$$\bar{\eta} = \eta/\varepsilon \tag{3.220}$$

Prandtl's boundary-layer equations are then obtained by applying the formal limit for $\varepsilon \downarrow 0$ to the resultant equations. If ξ and η are coordinates that are respectively tangential to and normal to the surface of the solid, $d\zeta$ and $d\eta$ provide the right length element (relative to x_i^*) at the surface, and u and v are the velocity components with respect to ξ and η, then the Prandtl equations are

$$\frac{\partial u}{\partial \xi} + \frac{\partial v}{\partial \eta} = 0, \tag{3.221a}$$

$$u\frac{\partial u}{\partial \xi} + v\frac{\partial u}{\partial \eta} + \frac{\partial p}{\partial \xi} = \frac{1}{R}\frac{\partial^2 u}{\partial \eta^2}, \tag{3.221b}$$

$$\partial p/\partial \eta = 0 \tag{3.221c}$$

as border conditions Prandtl suggested the obvious one, which was (3.212), and the more unexpected one. [27]

The values of p and u of the boundary-layer
solution at $\bar{\eta} = \infty$ equal the corresponding values
(denoted by P_e and U_e) of the Euler solution at $\eta = 0$.   (3.222)

We note that from the Euler equations it follows that

$$\frac{\partial}{\partial \xi} \left(\frac{u_e^2}{2} + p_e \right) = 0 \qquad (3.223a)$$

and, since a physically irrelevant constant may be added to the pressure,

$$u_e^2/2 + p_e = 0 \qquad (3.223b)$$

Detailed discussions of boundary-layer theory and higher order approximations from the perspective of asymptotic theory can be found, for example, in Lagerstrom (1964) and Van Dyke (1964). Besides, several specialised problems will be covered.

3.11.4   Role of sub characteristics

The characteristic surfaces (or lines) of the outer equations in a singular perturbation problem involving partial differential equations are referred to as subcharacteristics. The outer equations in our situation are the streamlines, or the lines that are always tangent to the velocity field, are the sole significant features of Euler equations. The boundary is a streamline as a result of (3.218), and on this streamline there is a discontinuity between the recommended values of the tangential velocity and those provided by the Euler solution. Latta (1951) demonstrated that the thickness of the boundary layer relies on whether the discontinuity occurs on a subcharacteristic for a variety of partial differential equations. Lagerstrom (1964) covers the application of this idea to fluid dynamics in great depth. Here is a simple example. The so-called Oseen equations emerge if the first two terms on the left side of (3.217b) (also known as the transport terms) are replaced for $\partial q_j / \partial x_1^*$. The lines $x_2^* =$const are now the characteristic lines. When the boundary is parallel to the x_1^*-axis, the thickness of the boundary layer is $R^{-1/2}$; it is R^{-1}- when the boundary has a positive slope; and there is no boundary layer when the boundary has a negative slope (loc. cit.). Cole (1968, Chap. 4) provides a general overview of the function of subcharacteristics.

---

[27]This corresponds to (3.83c). It was precisely to illustrate some basic aspects of Prandtl's perturbation method that Friedrichs (1953) introduced a model equation similar to our example in earlier section.

3.11.5   Nonuniqueness of the Euler solution

Actually, the asymptotic theory for flow at high Reynolds numbers gives rise to certain really challenging problems that have not yet been resolved. It is frequently challenging to choose the appropriate solution, that is, the one that is the limit of the Navier-Stokes equations as $R \to \infty$, because the solution of the Euler equations with the boundary conditions stated above is not unique.

A study of the vorticity can be helpful in some situations. We define

$$\text{vorticity} = \omega = \partial q_1 / \partial y - \partial q_2 / \partial x \tag{3.224}$$

It is then easily seen that vorticity obeys the equation

$$q_1 \frac{\partial \omega}{\partial x_1^*} + q_2 \frac{\partial \omega}{\partial x_2^*} = \frac{1}{R}\left(\frac{\partial^2 \omega}{\partial x_1^*} + \frac{\partial^2 \omega}{\partial y_2^*} \right). \tag{3.225}$$

A streamline is a curve that is tangent to $\mathbf{q}$ everywhere. As a result, equation (3.225) can be interpreted as follows: The streamlines move the vorticity while also causing it to diffuse with a coefficient of $1/R$ like heat. Every streamline has a constant vorticity when R is infinite. The boundary conditions call for $q_1 = 1$ and $q_2 = 0$, and hence $\omega = 0$ at $x_1 = \infty$, for flow past an object. If every streamline originates from upstream infinity, then vorticity is zero everywhere in the limit. By making this assumption, the Euler equations are reduced to the Laplace equation, which has a unique solution. It appears plausible that every streamline originates from upstream infinity for semi-infinite entities whose width increases monotonically downstream. The parabola with axis is the most basic illustration of such a body towards the upward-pointing x_1-axis. Van Dyke (1964a) calculated the two leading terms of the inner and outer expansions in this example

The limiting solution assumption for the case of flow past a finite body, however, results in contradictions with boundary layer theory (see Lagerstrom (1964)). Then, streamlines with nonzero vorticity must exist in the limiting flow; these streamlines must either be closed or originate from and return to infinity ($x_1 = \infty$) downstream. The so-called Kirchhoff flow, which separates a zone of potential flow from a region of zero velocity using two streamlines that originate at the solid and carry infinite vorticity, is a limiting example. However, no convincing justifications have yet been developed for characterising the limiting flow in a way that distinguishes the solution to the Euler equations. The challenges are related to the extremely difficult flow separation problem and the non-existence of Prandtl equation solutions, topics that cannot be discussed here.

We see that, in contrast to the examples given previously, the leading term of the outer expansion cannot be identified a priori, or put another way, we are unsure of what is perturbed. A solution would need simultaneous consideration of the inner and outer expansions' leading terms (and perhaps other expansions as well). Similar circumstances exist for the problem mentioned below, which is considerably easier to solve and can be done so with specific boundary conditions.

3.11.6   Limiting flow inside a finite domain

We take into account a finite domain D with a boundary B. We substitute (3.212) with in the boundary conditions for (3.211).

$$\text{normal velocity component} = 0 \qquad (3.226a)$$
$$\text{tangential velocity component} = \text{given function } u_b. \qquad (3.226b)$$

In the same domain, we employ (3.226a) and forego (3.226b) for the Euler flow. The solution is obviously not unique. If $(\mathbf{q}, p)$ is one solution, then $(k\mathbf{q}, k^2 p)$ with $k = \text{const}$ is another solution, and there are numerous other solutions that are drastically distinct. However, according to Prandtl (1905), the limiting flow has the following characteristic:

$$\text{The vorticity is constant in each vortex.} \qquad (3.227)$$

Two streamlines are said to be part of the same vortex if and only if they are completely enclosed by one another and the region between them consists entirely of these "nested" streamlines. Consequently, the flow shown in Fig.(3.3) is made up of two vortices. The enclosing streamlines are, respectively, ABC and ACD. Neither of the two vortices contain the streamline ABCD.

The assumption that streamlines in a finite domain are closed (which follows if one thinks that there are no mass sources or mass-sinks in the domain) is necessary for the proof of (3.227).

The two equations in (3.217b) can be written as

$$\frac{-\partial h}{\partial x_2^*} + q_1 \omega = \frac{1}{R} \frac{-\partial \omega}{\partial x_1^*} \qquad (3.228a)$$

$$\frac{\partial h}{\partial x_1^*} + q_2 \omega = \frac{1}{R} \frac{-\partial \omega}{\partial x_2^*} \qquad (3.228b)$$

where

$$h = \frac{1}{2}(q_1^2 + q_2^2) + p. \qquad (3.228c)$$

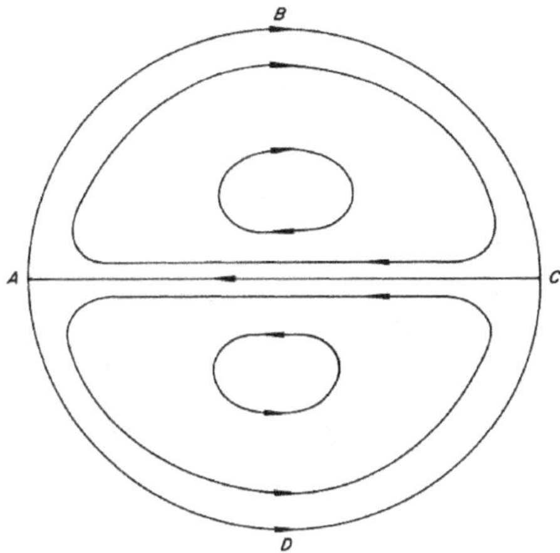

Figure 3.3: Vortices

A closed streamline S with the normal $dn = (-dx_2^*, dx_1^*)$ is considered. Integrate over the streamline after multiplying (3.228a), (3.228b) by dx_2^* and dx_1^*, respectively. Since h is one-valued, the integral of the first term, which calculates the increase in h as it circles a closed curved , is zero. Since $\mathbf{q} \cdot d\mathbf{n}$ is zero., the second term produces zero. Thus, we achieve

$$\oint_s \frac{\partial \omega}{\partial n} ds = 0, \tag{3.229}$$

where ds is the curve S's length component. This outcome is valid for arbitrary R. The stream function ψ is now introduced by

$$q_1 = \partial \psi / \partial x_2^*, \qquad q_2 = -\partial \psi / \partial x_1^*, \tag{3.230}$$

a definition that is made possible by (3.217a). Of course, the ψ value remains constant along a streamline. Along a streamline in the limit $R \to \infty$, the value of ω is constant , so that along each streamline $\omega = \omega(\psi)$ and furthermore $d\omega/d\psi$ are both constant . Then, we rewrite (3.229) as

$$\left(\frac{\partial \omega}{\partial \psi}\right)_s \oint_s \frac{\partial \psi}{\partial n} ds = 0 \tag{3.231}$$

The integral's result is nonzero because, according to (3.230), $\partial \psi / \partial n$ is the tangential velocity component. The conclusion assumes that S is a component of a vortex (it would be incorrect if S were a component of the streamline ABCD in 3.3). Hence

$$\partial \omega / \partial \psi = 0 \text{ for } R = \infty \tag{3.232}$$

This conclude the proof of (3.227).

Returning to the perturbation problem we consider first the flow inside a circle of radius unity in x_j^*-coordinates. The prescribed tangential velocity component u_b is now a function of the angle θ. We shall see if (3.227) will help in finding the solution. The obvious difficulty is that we do not know how many vortices there are. Assume tentatively that only one vortex occurs. It can easily be shown that the only solution of the Euler equations with constant vorticity is when the fluid rotates like a solid, i.e., when the r-component of the velocity is zero and the θ-component, denoted by u, is proportional to r. This still leaves a one parameter family of solutions: If (u, p) is a solution, then so is $(ku, k^2 p)$. The flow field is uniquely determined if u_e is given, where

$$u_e = u \text{ at } r = 1. \tag{3.233}$$

Thus we want to relate the unknown u_e to the known $u_b(\theta)$ through a boundary layer solution. We note that the *Euler flow* at $r = 1$ is u_e whereas the actual velocity field at $r = 1$ is u_b. We introduce the boundary-layer coordinates.

$$\xi = \theta, \qquad \eta = 1 - r, \qquad \bar{\eta} = \eta \sqrt{R} \tag{3.234}$$

As boundary conditions for the boundary-layer flow we then find

$$u = u_b \text{ at } \bar{\eta} = 0, \tag{3.235a}$$
$$u = u_e \text{ at } \bar{\eta} = \infty, \tag{3.235b}$$

The criterion for finding u_e is that the Navier-Stokes solution is periodic in θ and that hence the boundary-layer solution must be periodic in $\xi(= \theta)$. Our problem is then to translate this criterion into a priori estimate of u_e.

For the circle one proceeds as follows (Feynman and Lagerstrom (1956)).

We make the coordinate change from (ξ, η) to (ξ, ψ) (known to aerodynamicists as von Mises coordinates). The boundary-layer equation (3.221b) then becomes

$$u \frac{\partial u}{\partial \xi} + \frac{\partial p}{\partial \xi} = u \frac{\partial^2 (u^2 / 2)}{\partial \psi^2}. \tag{3.236}$$

This equation is valid when p_e is an arbitrary function of ξ. In the present case p_e is constant and (3.236) reduces to

$$2\frac{\partial u}{\partial \xi} = \frac{\partial^2 u^2}{\partial \xi^2}. \tag{3.237}$$

We define the average of a function $f(\xi, \psi)$ by

$$\bar{f}(\psi) = \frac{1}{2\pi} \int_0^{2\pi} f(\xi, \psi)d\xi, \qquad \psi = \text{const.}$$

We can now use periodicity: $u(\xi, \psi) = u(\xi + 2\pi, \psi)$ implies that the average of the left-hand side (of (3.237)) is zero. Then[28]

$$d^2\bar{u}^2/d\psi^2 = 0 \tag{3.238}$$

Thus the derivative of the mean of u^2 does not vary from streamline to streamline. Evaluation at $\bar{\eta} = \infty$ shows that this derivative actually is zero on every streamline. Thus $\bar{u}^2$ is the same on all streamlines. Evaluation at $\bar{\eta} = \infty$ and $\bar{\eta} = 0$ gives, finally,

$$u_e^2 = \bar{u}_b^2. \tag{3.239}$$

Hence we have found an a priori relation between the given function $u_b(\theta)$ and the unknown constant u_e.

Equation (3.239) does not give the sign of u_e. There is, however, a much more serious difficulty than that. Equation (3.239) is a necessary but not a sufficient condition for the existence of a boundary-layer solution. Consider the Euler flow as in Fig. 3.3 and assume that $\omega + 1$ in the upper half of the circle and $\omega = -1$ in the lower half of the circle. If one prescribes u_b to be equal to the values taken by the Euler flow on the boundary or deviating very slightly therefrom while still keeping the basic symmetry around AC, the limiting flow must be of the form shown in Fig. 3.3with possibly a slightly different value of $|\omega|$. Thus computation of a u_e from (3.239) is meaningless in this case.

The vortex picture shown in Fig. 3.3 is still relatively simple; there may be a large number of vortices in an asymmetrical arrangement.

At present the limiting vortex structure cannot be determined a priori. Thus, even for the flow inside a circle the problem is not solved. However, if the resulting flow has only one vortex, the solution may be obtained

---

[28]The argument used amounts to the consideration of momentum balance over one period.

with the aid of (3.239). The natural way to test whether there is one vortex is the following:

One computes $|u_e|$ by (3.239) and gives u_e the same sign as the average of u_b and studies the resulting boundary-layer equations. The breakdown of the boundary-layer solution indicates that there is more than one vortex. The theory of the boundary-layer equations furnishes approximate criteria for the nonexistence of a solution. Clearly, if u_b differs only slightly from the computed u_e, there will be only one vortex.

If the boundary is not a circle, the equivalent of (3.239) is not known. However, assume that the domain is simply connected. Let the given boundary condition u_b have the form

$$u_b = f(\xi) + \varepsilon g(\xi), \qquad (3.240)$$

where $f(\xi)$ is the value at the boundary of a special one-vortex Euler solution (which can be found by solving Poisson's equation). We assume ε to be small. Then the actual Euler flow corresponding to the boundary conditions (3.240) differs from the special one by a multiplicative constant k. Thus

$$u_e = k \cdot f. \qquad (3.241)$$

For $\varepsilon = 0, k = 1$. Hence

$$k = 1 + \varepsilon k_1 + \dots. \qquad (3.242)$$

The periodicity requirement then gives (Feynman and Lagerstrom (1956))

$$k_1 = \overline{f^2 g}/\overline{f^3} \qquad (3.243)$$

This result reduces to (3.239) when the boundary is a circle. It may be proved using von Mises coordinates. The proof is similar to that of (3.239).

3.11.7   Optimal coordinates

By using some clever geometrical reasoning, Kaplun (1954) demonstrated that if one can identify the first term of the inner expansion and the first two terms of the outer expansion for flow in two dimensions at high Reynolds numbers, one can then introduce a new system of coordinates known as optimal coordinates, where the leading term of the inner expansion contains the outer expansion's two leading terms. Legner (1971) expanded this work to a higher order. The two matching expansions have been replaced with one regular expansion up to the sequence taken into consideration.

3.11.8 Coordinate-type expansions, artificial parameters

The distinction between coordinates and parameters is not inherent, as was stated in §8; rather, it depends on the problem being examined and the approach taken to solve it. As the parameter ε goes to a limiting value, we sought for asymptotic expansions that are uniformly valid in a closed domain of coordinate space. The quantity ε was also a parameter in that the differential equations determining the function never included derivatives with respect to ε. There are instances where a variable behaves physically as a coordinate and appears in the differential equation like a coordinate, yet in the asymptotic solution we regard it as a parameter. Consider an expansion of $f(x_1, x_2, \varepsilon)$, for example, that is uniformly valid in x_2 as x_1 goes to infinity and for a fixed value of ε. As contrast to the parametertype expansions discussed before, these expansions are known as coordinate-type expansions; for more information, see Lagerstrom and Cole (1955) and Chang (1961).

Finding the flow field at a large distance from the body is a classic problem of coordinate-type expansions in the theory of viscous compressible flow past a finite body. The problem's two-dimensional case extensively examined by Chang (1961), who also covers the basic method for coordinate-type expansions. This problem is singuar; a different scale in x_2 is employed inside the narrowed wake zone known as the wake compared to outside the wake. Similar to a parameter-type expansion, the solutions for various regions are matched, and switchback even takes place. The quantity equal to the minor parameter ε is x_1^{-1}. By introducing an artificial dimensional parameter in the form of an artificial length, one can have the same formal outcome. The body dimension is then calculated as the small parameter by dividing it by the artificial length. The artificial length must, of course, be removed from the final answer (see Chang, 1961).

It must be emphasised that for a fixed Reynolds number, the flow near a body cannot equal the flow at great distances. The asymptotic expansion at vast distances will then contain unknown constants since the boundary constraints on the body determine the flow at large distances. Assume that another solution, such as a Taylor expansion in the proper coordinates, is discovered close to the body. Unknown constants will also be included in this expansion. But matching, or comparing partial sums, cannot relate the unknown constants of the two expansions. Only the full Taylor expansion close to the body will determine the asymptotic expansion at great distances, hence the problem is analytic continuation rather than matching. Integral conservation laws, on the other hand, can be thought of as a type of analytical continuation in the mean.[29] For instance, the integral of $\mathbf{q} \cdot d\mathbf{n}$ across any closed curve is zero because the divergence of the velocity field is zero (conservation of mass) and because the flow through the surface of the body is zero. This establishes one of the asymptotic expansion's constants at infinity. Similar to how other constants in this expansion might be connected to the overall force and torque the fluid exerts on the body.

Nonviscous flow past a parabola, or paraboloid, is another problem with a one length. One cannot speak of a slender parabola because the problem only has one independent real length, let's say the radius of curvature of the nose. Far downstream, the parabola appears slender, meaning that the ratio of the radius to the distance to the nose is small. By adding a significant artificial length and employing singular perturbation techniques, one can formalise this. Near the nose, the answer will not hold true, and matching is impossible. But in this case, the law of mass conservation can be quite useful. Information is provided in Van Dyke (1958).

We observe that, depending on the boundary circumstances, a physical parameter in a particular situation may be mathematically artificial. For instance, the viscosity is artificial in the sense that it needs to be removed in viscous flow past a semi-infinite plate (or any conical body, i.e., body without a length parameter). We can infer from dimensional analysis that v must only occur in combinations of x_j (dimensional coordinate). u_0/v. Since v will vanish from the equations (3.211) and the boundary conditions if we employ these coordinates. Of course, it is possible to expand for small values of v, but in practise, this is an expansion with a large x. At the nose of the flat plate, the expansion will not be uniformly valid. More specifically, it is simply demonstrated that if we have any approximation that is uniformly valid to order unity everywhere and where v occurs only in the combination described in the definition of x_j above, then the approximation is in fact the exact solution. For more information, see Lagerstrom and Cole (1955, pp. 869-870) and Chang (1961, pp. 821-822).

---

[29]This leads to the interesting question of finding all independent conservation laws for a system of equations. This problem is discussed in Casten (1970).

3.12   Appendix

The equation

$$\frac{d^2 f}{dx^2} + \frac{m}{x}\frac{df}{dx} + \frac{df}{dx} = 0 \tag{A.1a}$$

with the boundary condition

$$f(\infty) = 0 \tag{A.1b}$$

determines f to be a constant times the function

$$E_m(x) = \int_x^\infty e^{-t} t^{-m} dt. \tag{A.2}$$

The corresponding nonhomogeneous equations

$$\frac{d^2 f}{dx^2} + \frac{m}{x}\frac{df}{dx} + \frac{df}{dx} + E_m \frac{dE_m}{dx} = 0 \tag{A.3}$$

and

$$\frac{d^2 f}{dx^2} + \frac{m}{x}\frac{df}{dx} + \frac{df}{dx} + \left(\frac{dE_m}{dx}\right)^2 = 0 \tag{A.4}$$

have the particular solutions, with $f(\infty) = 0$,

$$f = F_m(x) = \frac{dE_{m-1}(x)}{dx} E_m(x) + 2^{2m-1} E_{2m-1}(2x) \tag{A.5}$$

and

$$f = G_m(x) = -\frac{1}{2}(E_m(x))^2,$$

respectively.

Asymptotic expansions valid for x small are

$$E_1(x) = -In\,x - \gamma + x + O(x^2) \tag{A.6a}$$

$$E_2(x) = \frac{e^{-x}}{2x^2} - E_1(x) = \frac{1}{x} + In\,x + (\gamma - 1) - \frac{x}{2} + \frac{x^2}{12} + O(x^3) \tag{A.6b}$$

$$E_3(x) = \frac{e^{-x}}{2x^2} - \frac{e^{-x}}{2x} + \frac{1}{2}E(x)$$

$$= \frac{1}{2x^2} - \frac{1}{x} - \frac{1}{2}In\,x + \frac{3 - 2\gamma}{4} + \frac{x}{6} + O(x^2)$$

$$F_1(x) = -In\,x - 2\,In\,2 - \gamma - x\,In\,x + (3 - \gamma)x + O(x^2\,In\,x) \tag{A.7a}$$

$$F_2(x) = -\frac{In\,x}{x} - \frac{\gamma+2}{x} - 3\,In\,x - 3\gamma = 5 - 4\,In\,2 + O(x\,In\,x) \quad \text{(A.7b)}$$

$$G_1 = -\frac{1}{2}In^2\,x - \gamma\,In\,x - \frac{\gamma^2}{2} + x\,In\,x + \gamma x + O(x^2) \quad \text{(A.8a)}$$

$$G_2 = -\frac{1}{2x^2} - \frac{1}{2}In\,x - \frac{\gamma-1}{x} - \frac{In^2\,x}{2} - (\gamma-1)\,In\,x\frac{-\gamma^2+2\gamma}{2}$$
$$+ \frac{x\,In\,x}{2} + \frac{\gamma-1}{2}x + O(x^2)$$

Chapter 4

Artificial Parameter Method

Perturbation techniques divide a nonlinear problem into an endless number of linear sub-problems using perturbation quantities and then estimate it by the sum of the first several sub-problems' answers. Although the existence of perturbation quantities is undoubtedly a foundational aspect of perturbation techniques, these techniques are severely constrained by the perturbation quantity. The existence of such a perturbation quantity in every nonlinear situation seems, to start, impossible. This clearly limits the use of perturbation techniques. Second, perturbation approximations are only appropriate for nonlinear situations with weak nonlinearity since analytical approximations of nonlinear problems frequently fail as nonlinearity grows stronger.

A few non-perturbative techniques exist. By introducing a so-called artificial small parameter, it is possible to avoid the dependence of perturbation techniques on small/large parameters. Lyapunov considered the following equation in 1892.

$$\frac{dx}{dt} = A(t)x,$$

$A(t)$ is a time periodic matrix in this case. In order to replace this equation with the equation given below, Lyapunov added an artificial parameter ϵ.

$$\frac{dx}{dt} = \epsilon A(t) x$$

and then calculated power series expansions over ϵ for the solutions. Lyapunov demonstrated that series converge for $\epsilon = 1$, therefore we can enter the final expression by setting $\epsilon = 1$. The methodology described above is known as Lyapunov's artificial small parameter method. Karmishin et al. expanded on this concept to propose the -expansion method. Karmishin et al. created an artificial parameter to replace the equation

$$x^5 + x = 1 \tag{4.1}$$

with the equation

$$x^{1+\delta} + x = 1 \tag{4.2}$$

and then computed power series expansions over δ to obtain the approximations by converging the series to $[3, 3]$ Padé approximants and setting $\delta = 4$. Essentially, the δ-expansion method is the same as Lyapunov's artificial small parameter method. It should be observed that both methods introduce an artificial parameter, although in different places and with different symbols in a given nonlinear equation. We also have a lot of freedom in replacing Equation (4.1) with other equations, such as

$$\delta x^5 + x = 1 \tag{4.3}$$

As Karmishin et al. point out, the approximation provided by the preceding equation is significantly worse than that provided by Equation (4.2). Both the artificial small parameter approach and the δ-expansion method require some basic guidelines to establish where the artificial parameter ϵ or δ should occur. Both the artificial small parameter method and the δ-expansion method, like perturbation techniques, do not give a convenient way to adjust the area of convergence and rate of approximation series.

Adomian's decomposition method is an effective analytic technique for solving severely nonlinear problems. Adomian's decomposition approach is applicable to ordinary and partial differential equations with small/large parameters, making it quite general. Furthermore, the Adomian approximation series converges quickly. Adomian's decomposition method, on the other hand, has certain limitations. Polynomials are frequently found in the approximate solutions provided by Adomian's decomposition method. In general, power series convergence regions are small, hence accelerating techniques are frequently required to extend convergence regions. This is mostly due to the fact that power series are not always an effective set of basis functions for approximating a nonlinear problem, but Adomain's decomposition method does not allow us to use alternate base functions. Adomian's decomposition approach, like the artificial small parameter method and the δ-expansion method, does not give a convenient approach for adjusting the convergence region and rate of approximation solutions.

4.1 Artificial Parameter Method

Artificial Parameter Method was introduced by Liao in 1997 which is not based on small parameter but depends on an Artificial Parameter.

$$A(u(t)) - f(r) = 0, \qquad r \in \Omega,$$

$$B\left(u, \frac{\partial u}{\partial n}\right) = 0, \qquad r \in \Gamma.$$

$$A = L + N$$

$$L(u(r)) + N(u(r)) - f(r) = 0. \tag{4.4}$$

$$L(u(r)) + pN(u(r)) - f(r) = 0. \tag{4.5}$$

Homotopy equation with an artificial Parameter.
In this method the solution is considered as power series;

$$u = u_0 + u_1 p + u_2 p^2 + \dots$$

if $p \to 1$
Equation (4.5) changes to (4.4), and also it's solution.

Example:

$$\frac{du(t)}{dt} + u^2(t) = 1, \qquad u(0) = 0.$$

with the exact solution

$$u(t) = \frac{1 - e^{-2t}}{1 + e^{-2t}}$$

To impose the artificial parameter, there are three possibilities;

$$\text{APM 1:} \qquad \frac{du(t)}{dt} + pu^2(t) = 1, \qquad u(0)0.$$

$$\text{APM 2:} \qquad \frac{du(t)}{dt} - [1 - pu(t)][1 + u(t)] = 0,$$

$$\text{APM 3:} \qquad \frac{du(t)}{dt} - [1 - u(t)][1 + pu(t)] = 0$$

By substitution of

$$u(t) = \sum_{i=0}^{2} u_i(t)p^i$$

in APM 1 considering initial condition and equating the terms with the same powers of p

$$\frac{du(t)}{dt} + pu^2(t) = 1, \qquad u(0) = 0.$$

$$p^0: \quad \frac{du_0}{dt} - 1 = 0, \qquad u_0(0)0,$$

$$p^1: \quad \frac{du_1}{dt} + u_0^2 = 0, \qquad u_1(0)0,$$

$$p^2: \quad \frac{du_2}{dt} + 2u_0 u_1 = 0, \qquad u_2(0)0,$$

$$\vdots$$

$$u_0(t) = t, \qquad u_1(t) = -\frac{1}{3}t^2, \qquad u_2(t) = \frac{2}{5}t^5,$$

$$u_{\text{APM 1}}(t) = t - \frac{1}{3}t^3 + \frac{2}{5}t^5,$$

$$\frac{du(t)}{dt} + pu^2(t) = 1, \qquad u(0) = 0.$$

$$p^0: \quad \frac{du_0}{dt} - 1 = 0, \qquad\qquad u_0(0) = 0,$$

$$p^1: \quad \frac{du_1}{dt} + u_0^2 = 0, \qquad\qquad u_1(0) = 0,$$

$$p^2: \quad \frac{du_2}{dt} + 2u_0 u_1 = 0, \qquad\qquad u_2(0) = 0,$$

$$\vdots$$

$$u_0(t) = t, \qquad u_1(t) = -\frac{1}{3}t^2, \qquad u_2(t) = \frac{2}{5}t^5,$$

$$u_{\text{APM 1}(t)} = t - \frac{1}{3}t^3 + \frac{2}{5}t^5,$$

$$\frac{du(t)}{dt} - [1 - pu(t)][1 + u(t)] = 0,$$

$$p^0: \qquad \frac{du_0}{dt} - u_0 - 1 = 0, \qquad\qquad u_0(0) = 0,$$

$$p^1: \qquad \frac{du_1}{dt} + u_0 + u_0^2 - u_1 = 0, \qquad\qquad u_1(0) = 0,$$

$$p^2: \qquad \frac{du_2}{dt} + 2u_0 u_1 = 0, \qquad\qquad u_2(0) = 0,$$

$$\vdots$$

$$u_0(t) = e^t - 1, \qquad u_1(t)(t - e^t + a)e^t,$$

$$u_2(t) = (\frac{1}{2}t^2 - e^t + t - 2te^t + e^{2t})e^t$$

$$u_{\text{APM } 2}(t) = e^t - 1 + (t - e^t + 1)e^t + (\frac{1}{2}t^2 - e^t + t - 2te^t + e^{2t})e^t$$

$$\frac{du(t)}{dt} - [1 - u(t)][1 + pu(t)] = 0$$

$$p^0: \qquad \frac{du_0}{dt} - u_0 - 1 = 0, \qquad\qquad u_0(0) = 0,$$

$$p^1: \qquad \frac{du_1}{dt} + u_0 + u_0^2 + u_1 = 0, \qquad\qquad u_1(0) = 0,$$

$$p^2: \qquad \frac{du_2}{dt} - u_1 + 2u_0 u_1 + u_2 = 0, \qquad\qquad u_2(0) = 0,$$

$$\vdots$$

$$u_0(t) = 1 - e^t, \qquad u_1(t) = (t + e^{-t} - 1)e^{-t},$$

$$u_2(t) = (-\frac{1}{2}t^2 + e^{-t} + t - 2te^{-t} - e^{2t})e^{-t}$$

$$u_{\text{APM } 3}(t) = 1 - e^{-t} + (t + e^{-t} - 1)e^{-t} + (-\frac{1}{2}t^2 + e^{-t} + t - 2te^{-t} - e^{2t})e^{-t}$$

Table 4.1: Result obtained by artificial parameter method (APM-1)

t	$u_{\text{APM 1}}$	u_{exact}	$\lvert u_{\text{exact}} - u_{\text{APM 1}} \rvert$
0.1	0.0996680000	0.0966799462	2.9880538×10^{-3}
0.2	0.1973760000	0.1973753203	6.797×10^{-7}
0.3	0.2913240000	0.2913126125	1.13875×10^{-5}
0.4	0.3800320000	0.3799489623	8.30377×10^{-5}
0.5	0.4625000000	0.4621171573	3.828427×10^{-4}
0.6	0.5383680000	0.5370495670	1.318433×10^{-3}
0.7	0.6080760000	0.6043677771	3.7082229×10^{-3}
0.8	0.6730240000	0.6640367703	8.9872297×10^{-3}
0.9	0.7357320000	0.7162978704	1.9434129×10^{-2}
1	0.8000000000	0.7615941561	3.8405843×10^{-2}

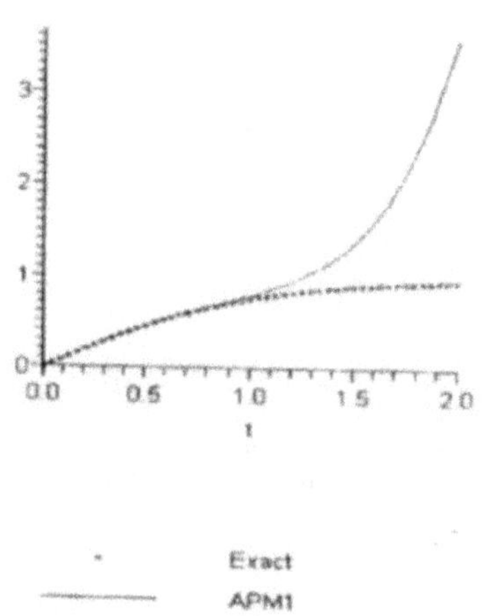

Figure 4.1: Result obtained by artificial parameter method (APM-1)

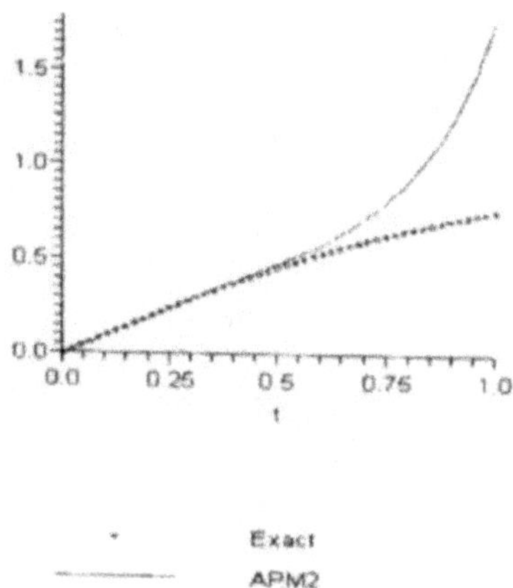

Figure 4.2: Result obtained by artificial parameter method (APM-2)

Table 4.2: Result obtained by artificial parameter method (APM-2)

t	$u_{\text{APM 1}}$	u_{exact}	$\left|u_{\text{exact}} - u_{\text{APM 1}}\right|$
0.1	0.0996746135	0.0966799462	2.9944673×10^{-3}
0.2	0.1975342018	0.1973753203	1.588815×10^{-4}
0.3	0.2924707743	0.2913126125	1.1581618×10^{-3}
0.4	0.3850574508	0.3799489623	5.1084885×10^{-3}
0.5	0.4790975540	0.4621171573	1.6980396×10^{-2}
0.6	0.5840348594	0.5370495670	4.6985292×10^{-2}
0.7	0.7186186468	0.6043677771	$1.14250869 \times 10^{-1}$
0.8	0.9163800969	0.6640367703	$2.52343326 \times 10^{-1}$
0.9	1.233702442	0.7162978704	$5.17404571 \times 10^{-1}$
1	0.761580761	0.7615941561	1.33951×10^{-5}

Table 4.3: Result obtained by artificial parameter method (APM-3)

t	$u_{\text{APM 1}}$	u_{exact}	$\lvert u_{\text{exact}} - u_{\text{APM 1}}\rvert$
0.1	0.0996655953	0.0966799462	2.9856491×10^{-3}
0.2	0.1973566176	0.1973753203	1.87027×10^{-5}
0.3	0.2912843022	0.2913126125	2.83103×10^{-5}
0.4	0.3799908861	0.3799489623	4.19238×10^{-5}
0.5	0.4624022891	0.4621171573	2.851318×10^{-4}
0.6	0.5378210778	0.5370495670	7.715108×10^{-4}
0.7	0.6058871685	0.6043677771	1.5193914×10^{-3}
0.8	0.6665237997	0.6640367703	2.4870294×10^{-3}
0.9	0.7198796209	0.7162978704	3.5817505×10^{-3}
1	0.7662732111	0.7615941561	4.679055×10^{-3}

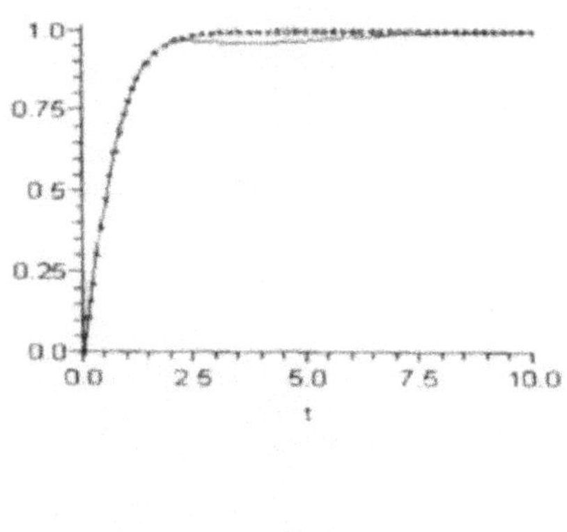

Figure 4.3: Result obtained by artificial parameter method (APM-3)

$$\frac{du(t)}{dt} + u^2(t) = 1$$

$$\frac{du(t)}{dt} + \beta u(t) + u^2(t) - \beta u(t) = 1,$$

$$u(t) = \sum_{i=0}^{2} v_i p^i$$

$$H(v,p) = (1-p)\left[\frac{\partial v}{\partial t} + \beta v - \frac{\partial u_0}{\partial t} - \beta u_0\right] + p\left[\frac{\partial v}{\partial t + v^2 - 1}\right] = 0$$

$$p \in [0,1], \qquad r \in \Omega.$$

$$H(v,0) = L(v) - L(u_0) = 0,$$
$$H(v,1) = A(v) - f(r) = 0.$$

Where p is a parameter.
By substituting

$$\sum_{i=0}^{2} v_i p^i$$

and equating the coefficient of the terms with the same Power of p

$$p^0: \qquad \frac{dv_0}{dt} + \beta v_0 - \frac{du_0}{dt} - \beta u_0 = 0, \qquad\qquad v_0(0) = 0,$$

$$p^1: \qquad \frac{dv_1}{dt} + \beta v_1 + v_0^2 + \frac{du_0}{dt} - \beta v_0 + \beta u_0 - 1 = 0, \qquad v_1(0) = 0,$$

from this relation we have

$$\frac{dv_1}{dt} + \beta v_1 + \left[\frac{dv_0}{dt} + v_0^2 - 1\right] = 0, \qquad v_1(0) = 0, \qquad (4.6)$$

Let's take $u_0(t) = 1 - e^{-\beta t}$ which satisfies initial condition
$v_0(0) = 1 - e^{-\beta t}, \quad v_0(0) = 0.$
Substituting $v_0(t)$ into (4.6)

$$\frac{dv_1}{dt} + \beta v_1 + \beta e^{-\beta t} + 1 + e^{-2\beta t} - 2e^{-\beta t} - 1 = 0,$$

$$\frac{dv_1}{dt} + \beta v_1 + (\beta - 2)e^{-\beta t} + e^{-2\beta t} = 0,$$

In the last equation let's take $\beta = 2$,

$$\frac{dv_1}{dt} + 2v_1(t) = -e^{-4t}$$

$$v_1(t) = ce^{-2t} + \frac{1}{2}e^{-4t}$$

$$v_1(0) = 0 \Rightarrow c = -\frac{1}{2}.$$

$$v_1(t) = \frac{1}{2}\left(e^{-4t} - e^{-2t}\right)$$

A two terms approximation to the solution is:

$$u(t) = 1 - e^{-2t} + \frac{1}{2}\left(e^{-4t} - e^{-2t}\right)$$

4.1.1   An illustrative example

Think about a free sphere dropping from a static state and falling in the air. Let $\tilde{t}$ denote the time, $U(\tilde{t})$ the velocity of the sphere, m the mass, and g the acceleration of gravity. Assume that the air resistance on the sphere is a $U^2(\tilde{t})$, where a is a constant. Then, due to second law of Newton, it holds

$$m\frac{dU(\tilde{t})}{d\tilde{t}} = mg - aU^2(\tilde{t}) \tag{4.7}$$

subject to initial condition

$$U(0) = 0. \tag{4.8}$$

In terms of physics, gravity causes a freely falling sphere's speed to increase until a constant velocity U_∞ is attained. Therefore, even without detailed knowledge of the solution $U(\tilde{t})$, we may get the limit velocity U_∞ directly from (4.7), i.e.,

$$U_\infty = \sqrt{\frac{mg}{a}}. \tag{4.9}$$

Writing down U_∞ as the characteristic velocity and U_∞/g as the characteristic time

$$\tilde{t} = \left(\frac{U_\infty}{g}\right)t, \qquad U(\tilde{t}) = U_\infty V(t), \tag{4.10}$$

we have the dimensionless equation

$$\dot{V}(t) + V^2(t) = 1, \qquad t \geq 0, \tag{4.11}$$

subject to the initial condition

$$V(0) = 0, \tag{4.12}$$

where the dot represents the derivative with respect to t and t symbolises the dimensionless time. As $t \to +\infty$, i.e., $\tilde{t} \to \infty$ and $U(\tilde{t}) \to U_\infty$, it follows from (4.10) that

$$\lim_{t\to+\infty} V(t) = 1, \tag{4.13}$$

even without solving Equation (4.11) and (4.12).

The exact solution of equation (4.11) and (4.12) is

$$V(t) = \tanh(t), \tag{4.14}$$

useful to the comparison of different approximations.

4.2 Solution given by some previous analytic techniques

In order to make a comparison, we first solve the example of nonlinear problem using some well-known analytical techniques.

4.2.1 Perturbation method

Assuming that the dimensionless time t is a small variable (referred to as the perturbation quantity), we first define $V(t)$ in a power series to get the perturbation approximation.

$$V(t) = \alpha_0 + \alpha_1 t + \alpha_2 t^2 + \alpha_3 t^3 + \ldots, \tag{4.15}$$

Using the initial condition (4.12) we gain $\alpha_0 = 0$. Then , substituting the above expression into equation (4.11), we obtain

$$\sum_{k=0}^{+\infty} \left[(k+1)\alpha_{k+1} + \sum_{j=0}^{k} \alpha_j \alpha_{k-j} \right] t^k = 1,$$

which holds for any $t \geq 0$, provided
$\alpha_1 = 1$,
$$\alpha_{k+1} = -\frac{1}{k+1} \sum_{j=0}^{k} \alpha_j \alpha_{k-j}, \qquad k \geq 1,$$
we therefore have the perturbation solution

$$V_{pert}(t) = t - \frac{1}{3}t^3 + \frac{2}{15}t^5 - \frac{17}{315}t^7 + \cdots = \sum_{n=0}^{+\infty} \alpha_{2n+1} t^{2n+1}, \tag{4.16}$$

which converges in a rather small region $0 \leq t < \rho_0$, where $\rho_0 \approx 3/2$, as shown in Figure (4.2.1). Note that the convergence region and rate of the perturbation solution (4.16) are uniquely determined.

4.2.2 Lyapunov's artificial small parameter method

Using the artificial small parameter approach proposed by Lyapunov, we first replace Equation (4.11) for Equation

$$\dot{V}(t) = +\varepsilon V^2(t) = 1 \tag{4.17}$$

and then write

$$V(t) = V_0(t) + \varepsilon V_1(t) + \varepsilon^2 V_2(t) + \ldots, \tag{4.18}$$

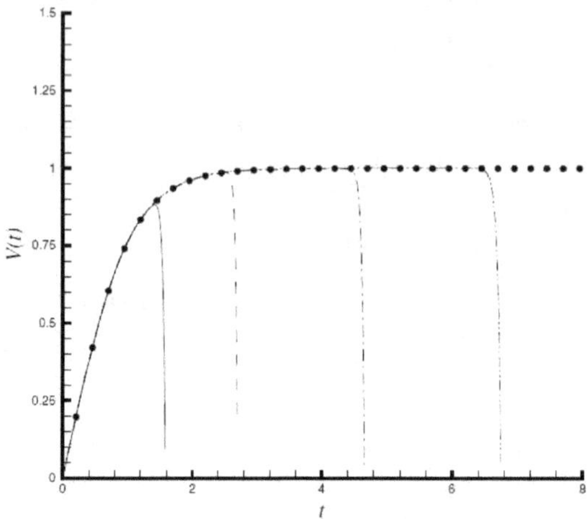

Figure 4.4: Comparison of the exact solution (4.14) with the solution expression (4.68). Symbols: exact solution; solid line: perturbation solution (4.16); dashed line: solution (4.68) when $\hbar = -1/2$; dash-dotted line: solution (4.68) when $\hbar = -1/5$; dash-dot-dotted line: solution (4.68) when $\hbar = -1/10$.

where an artificial small parameter ε is used. Equations (4.17) and (4.12) are obtained by substituting Equation (4.18) into them, and then balancing the coefficients of the power series of ε., we obtain

$$\dot{V}_0(t) = 1, \qquad\qquad V_0(0) = 0,$$
$$\dot{V}_1(t) + V_0^2(t) = 0 \qquad\qquad V_1(0) = 0,$$

$$\vdots$$

which give

$$V_0(t) = t, \qquad V_1(t) = -\frac{t^3}{3}, \qquad V_2(t) = \frac{2t^5}{15}, \qquad \cdots,$$

successively. Finally, setting $\varepsilon = 1$ in Equation (4.18), we have

$$V(t) = t - \frac{1}{3}t^3 + \frac{2}{15}t^5 - \frac{17}{315}t^7 + \cdots = \sum_{n=0}^{+\infty} \alpha_{2n+1} t^{2n+1}, \qquad (4.19)$$

This, as shown in Figure (4.2.1), is exactly the same as the perturbation solution (4.16) and is hence valid in a rather restricted region of t.

Note that the solution (4.19), obtained using Lyapunov's artificial small parameter approach, has a unique convergence region and rate.

4.2.3   Adomian's decomposition method

We first replace equations (4.11) and (4.12) using Adomian's decomposition method, we first replace Equations by

$$V(t) = t - \int_0^t V^2(t)dt, \qquad (4.20)$$

The solution is given by

$$V(t) = V_0(t) + \sum_{k=0}^{+\infty} V_k(t),$$

where

$$V_0(t) = t,$$

$$V_k(t) = -\int_0^t A_{k-1}(t)dt, \qquad k \geq 1$$

and

$$A_k(t) = \sum_{n=0}^{k} V_n(t)V_{k-n}(t)$$

is what's known as the Adomian polynomial. We gain in steps.

$$V_1(t) = -\frac{t^3}{3}, \qquad V_2(t) = \frac{2t^5}{15}, \qquad V_3(t) = \frac{17}{315}t^7, \ldots$$

such that

$$V(t) = t - \frac{1}{3}t^3 + \frac{2}{15}t^5 - \frac{17}{315}t^7 + \cdots = \sum_{n=0}^{+\infty} \alpha_{2n+1}t^{2n+1}, \qquad (4.21)$$

This, as shown in Figure (4.2.1), is exactly the same as the perturbation solution (4.16) and is hence valid in a rather restricted region of t. Note that the solution (4.21), obtained using Lyapunov's artificial small parameter approach, has a unique convergence region and rate.

4.2.4   The δ-expansion method

We first replace equation (4.11) by using the δ -expansion method

$$\dot{V}(t) + V^{1+\delta}(t) = 1, \tag{4.22}$$

where δ is a real number. Write

$$V(t) = V_0(t) + \sum_{n=1}^{+\infty} V^n(t)\delta^n. \tag{4.23}$$

Then, in the power series of δ, we expand $V^{1+\delta}(t)$ as follows.

$$V^{1+\delta} = V_0 + [V_1 + V_0 \, In \, V_0]\delta +$$
$$\left[(V_1(1 + In \, V_0) + \frac{1}{2} V_0 \, In^2 \, V_0 + V_2) \right] \delta^2 + \dots \tag{4.24}$$

The following equations are obtained by substituting Equations (4.23) and (4.24) into Equation (4.22) and balancing the power series of δ:

$$\dot{V}_0 + V_0 = 1, \qquad V_0(0) = 0$$
$$\dot{V}_1 + V_1 = -V_0 \, In \, V_0, \qquad V_1(0) = 0$$
$$\dot{V}_2 + V_2 = -V_1(1 + In \, V_0) - \frac{1}{2} V_0 \, In^2 V_0, \qquad V_2(0) = 0$$
$$\dot{V}_3 + V_3 = -V_2(1 + In \, V_0) - V_1 \left(1 + \frac{1}{2} In \, V_0 \right) In^2 V_0$$
$$\qquad - \frac{1}{6} V_0 \, In^3 \, V_0 - \frac{V_1^2}{2V_0}, \qquad V_3(0) = 0,$$
$$\vdots$$

By solving the aforementioned linear equations successively, we arrive at

$$V_0(t) = 1 - exp(-t),$$
$$V_1(t) = exp(-t) \left[t - \frac{\pi^2}{6} + P_2^L(e^{-t}) \right] - (1 - e^{-t})In(1 - e^{-t}),$$
$$\vdots$$

where

$$P_n^L(z) = \sum_{k=1}^{+\infty} \frac{z^k}{k^n}$$

is the n th polylogarithm function of z. The first-order approximation is

$$V(t) \approx 1 + exp(-t)\left[t - \frac{\pi^2}{6} - 1 + P_2^L(e^{-t})\right] - (1 - e^{-t})In(1 - e^{-t})..$$

$$(4.25)$$

The initial approximations of this type of solution appear to be valid throughout the entire region $0 \le t < +\infty$. Nevertheless, getting higher-order approximations gets harder and harder when the special function $P_n^L(z)$ appears.

It is interesting that the solutions obtained for the example problem using the perturbation approach, Lyapunov's artificial small parameter method, and Adomian's decomposition method are identical. The valid range for this solution, however, is only $0 \le t < 3/2$. This demonstrates how solutions provided by earlier non-perturbation techniques like Lyapunov's artificial small parameter method and Adomian's decomposition method may fail as physical parameters or variables rise and non-linearity becomes more pronounced, just like perturbation approximations. This fact also suggests the possibility of certain relationships between them. It should be highlighted that all of these earlier analytic methods' convergence regions and rates of solutions are uniquely determined. Neither perturbation techniques nor earlier non-perturbation methods like Lyapunov's artificial small parameter method, Adomian's decomposition method, or the δ-expansion method can alter these results or give us a convenient means to control and adjust the rate and region of convergence of solution series. Furthermore, it appears that none of these earlier analytical methods are useful for the significant property (4.13) of $V(t)$ at infinity, which we discover without resolving the problem. This example shows that, in essence, neither perturbation approaches nor the earlier non-perturbation methods can give us any means to utilize these kinds of relevant information to more accurately simulate a specific nonlinear problem. Finally, keep in mind that while the artificial parameters and δ appear in various places in Equations (4.17) and (4.22), respectively, the solution (4.25) provided by the δ-expansion method is valid throughout the entire region $0 \le t < +\infty$ and is therefore significantly superior to the solution (4.19) provided by Lyapunov's artificial small parameter method. If we define ε and δ as the embedding parameter, Equations (4.17) and (4.22) can be seen as a type of homotopy, as was described earlier. This example shows that it seems crucial to correctly create a homotopy in order to more effectively mimic a particular nonlinear situation. But up until now, we haven't had any fundamental rules to direct us.

4.3   Homotopy Analysis Method

By using homotopy, a fundamental notion of topology, a homotopy analysis method was proposed. The homotopy concept is quite simple and straightforward. Consider a differential equation as an example

$$\mathcal{A}[u(t)] = 0, \tag{4.26}$$

When $\mathcal{A}$ is a nonlinear operator, t represents time, and $u(t)$ is an unknown variable. Let $u_0(t)$ represent an initial approximation to $u(t)$, and $\mathcal{L}$ represent an auxiliary linear operator with the property

$$\mathcal{L}f = 0, \qquad \text{when } f = 0. \tag{4.27}$$

Then we build the so-called homotopy.

$$\mathcal{H}[\phi(t;q);q] = (1-q)\mathcal{L}[\phi(t;q) - u_0(t)] + q\mathcal{A}[\phi(t;q)], \tag{4.28}$$

where $\phi(t;q)$ is a function of t and q and $q \in [0,1]$ is an embedding parameter. We have when $q = 0$ and when $q = 1$.

$$\mathcal{H}[\phi(t;q);q]|_{q=0} = \mathcal{L}[\phi(t;0) - u_0(t)]$$

and

$$\mathcal{H}[\phi(t;q);q]|_{q=1} = \mathcal{A}[\phi(t;1)],$$

respectively. It is evident from (4.27) that

$$\phi(t;0) = u_0(t)$$

is the equation's solution

$$\mathcal{H}[\phi(t;q);q]|_{q=0} = 0$$

and

$$\phi(t;1) = u(t)$$

As a result, the equation's answer is evident.

$$\mathcal{H}[\phi(t;q);q]|_{q=1} = 0$$

As the embedding parameter q increases from 0 to 1, the solution $\phi(t; q)$ of the equation

$$\mathcal{H}[\phi(t; q); q] = 0$$

varies from the initial approximation $u_0(t)$ to the solution $u(t)$ of Equation (4.26) and is dependent on the embedding parameter q. In topology, this type of continuous variation is known as deformation.

Some numerical techniques, such as the continuation method and the homotopy continuation method, were created based on the concept of homotopy. In reality, the homotopy can represent the artificial small parameter technique and the δ-expansion method if we replace the artificial parameter ϵ or δ by the embedding parameter q. However, while the above-mentioned traditional method of constructing the homotopy (4.26) may be enough from the standpoint of numerical techniques, it is insufficient from the standpoint of analytic techniques. This is mostly due to the fact that we have a great deal of freedom in selecting the so-called auxiliary operator $\mathcal{L}$ and the initial approximations but lack of any principles to guide their selection. More crucially, the traditional method of constructing a homotopy does not allow for the easy adjustment of the convergence region and rate of approximation series.

Definition 4.3.1. *A topology on set X is a family, say τ of subset of X such that:*

a *The Union of elements of any sub family of τ belongs to τ.*

b *Intersection of elements of any subfamily of τ belongs to τ*

A topology space is denoted by (X, τ) where τ is a topology on X

Example: Let X be any set, and τ the family of all subsets of X i.e., $\tau = P(X)$ then τ is a topology on X which is called the discrete or trivial topology.

In topology, two continuous functions from one topological space to another are called homotopic, if one can be "continuously deformed" into another. Such approximation is known as homotopy between two functions. For example; IF X and Y are two topological space and there are two continuous mappings $f, g : X \to Y$ then $f(x)$ and $g(x)$ are called homotopic if there is a continuous function $H : X \times I \to Y$ such that $H(x, 0) = f(x)$ and $H(x, 1) = g(x)$, where $I = [0, 1]$. H is called homotopy.

$f \approx g$ means that f, and g are homotopic.

Lemma 4.3.1. *If A and B are two closed subset of X, such that $X = A \bigcup B$ and $f : A \to Y$, $g : B \to Y$ are two continuous functions. If for all $x \in A \bigcap B$, $f(x) = g(x)$ then the function $h : X \to Y$ defined as the following, is continuous.*

$$h(x) = \begin{cases} f(x) & x \in A, \\ g(x) & x \in B. \end{cases}$$

Example: if X is a topological space and the functions $f, g : X \to \mathfrak{R}^2$ are continonuous, then $f \approx g$. Since H by the following definition is a homotopy.

$$H : X \times I \to \mathfrak{R}^2,$$

$$H(x, t) = (1 - t)f(x) + tg(x).$$

4.4   Homotopy analysis solution

The essential notions of the homotopy analysis method are covered in this section using the same illustrative problem as in the previous section.

4.4.1   Zero-order deformation equation

Let $V_0(t)$ represent an initial guess of $V(t)$ which satisfies the initial condition (4.12), i.e.,

$$V_0(t) = 0 \tag{4.29}$$

Let $q \in [0, 1]$ represent the embedding parameter. The homotopy analysis approach is based on a continuous mapping $V(t) \to \Phi(t; q)$, that varies from the initial guess $V_0(t)$ to the actual answer $V(t)$ as the embedding parameter q increases from 0 to 1. To ensure this, use an auxiliary linear operator such as

$$\mathcal{L}[\Phi(t; q)] = \gamma_1(t)\frac{\partial \Phi(t; q)}{\partial t} + \gamma_2(t)\Phi(t; q), \tag{4.30}$$

where $\gamma_1(t) \neq 0$ and $\gamma_2(t)$ are real functions that will be determined later. The nonlinear operator is defined by Equation (4.11).

$$\mathcal{N}[\Phi(t; q)] = \frac{\partial \Phi(t; q)}{\partial t} + \Phi^2(t; q) - 1, \tag{4.31}$$

Let $\hbar \neq 0$ and $H(t) \neq 0$ represent the so-called auxiliary parameter and auxiliary function, respectively. We create a family of equations using the embedding parameter $q \in [0, 1]$.

$$(1 - q)\mathcal{L}[\Phi(t; q) - V_0(t)] = \hbar q H(t) \mathcal{N}[\Phi(t; q)], \qquad (4.32)$$

subject to the initial condition

$$\Phi(0; q) = 0. \qquad (4.33)$$

It should be noted that we have complete control over the auxiliary parameter $\hbar$, auxiliary function $H(t)$, initial approximation $V_0(t)$, and auxiliary linear operator $\mathcal{L}$. As demonstrated later in this book, such freedom plays key role in establishing the validity and flexibility of the homotopy analysis method.

Equation(4.32) becomes when $q = 0$.

$$\mathcal{L}[\Phi(t; 0) - V_0(t)] = 0, \qquad\qquad t \geq 0, \qquad (4.34)$$

subject to the initial condition

$$\Phi(0; 0) = 0. \qquad (4.35)$$

According to Equations (4.29) and (4.30), the solution of Equations (4.34) and (4.35) is simply

$$\Phi(t; 0) = V_0(t). \qquad (4.36)$$

When $q = 1$, Equation (4.32) becomes

$$\hbar H(t) \mathcal{N}[\Phi(t; 1)] = 0, \qquad t \geq 0, \qquad (4.37)$$

subject to the initial condition

$$\Phi(0; 1) = 0. \qquad (4.38)$$

Since $\hbar \neq 0, H(t) \neq 0$, and using the definition (4.31), Equations (4.37) and (4.38) are equivalent to the original equations (4.11) and (4.12), assuming

$$\Phi(t; 1) = V(t). \qquad (4.39)$$

As a result of Equations (4.36) and (4.39), $\Phi(t; q)$ varies as the embedding parameter q grows from 0 to 1 from the initial guess $V_0(t)$ to the exact answer $V(t)$. This type of variation is known as deformation in topology, and Equations (4.32) and (4.33) construct the homotopy $\Phi(t; q)$.

Equations (4.32) and (4.33) are referred to as the zero-order deformation equations. For brevity.

Equations (4.32) and (4.33) are called the *zero-order deformation equations.*

Because we have the freedom to choose the auxiliary parameter $\hbar$, the auxiliary function $H(t)$, the initial approximation $V_0(t)$, and the auxiliary linear operator $\mathcal{L}$, we can assume that they are all properly chosen so that the solution $\Phi(t; q)$ of the zero-order deformation equations (4.32) and (4.33) exists for $0 \leq q \leq 1$, and besides its mth-order derivative with respect to the embedding parameter q, i.e.,

$$V_0^{[m]}(t) = \frac{\partial^m \Phi(t; q)}{\partial q^m}\Big|_{q=0} \tag{4.40}$$

exists, where $m = 1, 2, 3, \ldots$. For brevity, $V_0^{[m]}(t)$ is called the *mth-order deformation derivative.* Define

$$V_m(t) = \frac{V_0^{[m]}(t)}{m!} = \frac{1}{m!}\frac{\partial^m \Phi(t; q)}{\partial q^m}\Big|_{q=0} \tag{4.41}$$

By Taylor's theorem, we expand $\Phi(t; q)$ in a power series of the embedding parameter q as follows:

$$\Phi(t; q) = \Phi(t; 0) + \sum_{m=1}^{+\infty} \frac{1}{m!}\frac{\partial^m \Phi(t; q)}{\partial q^m}\Big|_{q=0} q^m \tag{4.42}$$

From Equations (4.36) and (4.41), the above power series becomes

$$\Phi(t; q) = V_0(t) + \sum_{m=1}^{+\infty} V_m(t)\, q^m \tag{4.43}$$

Assume that the auxiliary parameter $\hbar$, auxiliary function $H(t)$, initial approximation $V_0(t)$, and auxiliary linear operator $\mathcal{L}$ are all selected in such a way that the series (4.43) converges at $q = 1$. Then, at $q = 1$, the series (4.43) becomes

$$\Phi(t; 1) = V_0(t) + \sum_{m=1}^{+\infty} V_m(t) \tag{4.44}$$

Therefore, using Equation (4.39), we have

$$V(t) = V_0(t) + \sum_{m=1}^{+\infty} V_m(t) \tag{4.45}$$

The preceding expression establishes a relationship between the original guess $V_0(t)$ and the exact solution $V(t)$ using the terms $V_m(t)(m = 1, 2, 3, \ldots)$, which were previously unknown.

4.4.2 High-order deformation equation

Define the vector

$$\vec{V}_n = \{V_0(t), V_1(t), V_2(t), \ldots, V_n(t)\}$$

The governing equation and accompanying initial condition of $V_m(t)$ can be determined from the zero-order deformation equations (4.32) and (4.33), according to definition (4.41). We get the mth-order deformation equation by differentiating Equations (4.32) and (4.33) m times with respect to the embedding parameter q, then setting $q = 0$ and lastly dividing them by $m!$

$$\mathcal{L}[V_m(t) - \chi_m V_{m-1}(t)] = \hbar H(t) R_m(\vec{V}_{m-1}), \qquad (4.46)$$

depending on the initial condition

$$V_m(0) = 0, \qquad (4.47)$$

where

$$R_m(\vec{V}_{m-1}) = \frac{1}{(m-1)!} \frac{\partial^{m-1} \mathcal{N}[\Phi(t; q)]}{\partial q^{m-1}} \Big|_{q=0} \qquad (4.48)$$

and

$$\chi_m = \begin{cases} 0 & \text{when} \quad m \leq 1 \\ 1 & \text{otherwise} \end{cases} \qquad (4.49)$$

From Equations (4.31) and (4.48), we have

$$R_m(\vec{V}_{m-1}) = \dot{V}_{m-1}(t) + \sum_{j=0}^{m-1} V_j(t)V_{m-1-j}(t) - (1 - \chi_m). \qquad (4.50)$$

Notice that $R_m(\vec{V}_{m-1})$ given by the above expression is only dependent upon

$$V_0(t), V_1(t), V_2(t), \ldots, V_{m-1}(t),$$

This can be found by solving the mth-order deformation equations (4.46) and (4.47). Thus, Equation (4.46) is a linear first-order differential equation subject to the linear initial condition (4.47), according to the specification (4.30) of the auxiliary operator $\mathcal{L}$. As a result, the solution $V_m(t)$ of high-order deformation equations (4.46) and (4.47) is simple to obtain, particularly using calculation tools such as Mathematica, Maple, MathLab, and so on. According to (4.45), we basically divide the original nonlinear problem, which is governed by Equations (4.11) and (4.12), into an infinite number of linear subproblems, which are governed by high-order deformation equations (4.46) and (4.47), and then approximate the precise solution using the sum of the solutions $V_m(t)$ of the first several sub-problems. It should be noted that such a transformation does not need the presence of any small or large parameters in the governing equation or initial/boundary conditions.

$V(t)$'s mth-order approximation is provided by

$$V(t) \approx \sum_{n=0}^{m} V_n(t). \tag{4.51}$$

It should be noted that the zero-order deformation equation (4.32) is determined by the auxiliary linear operator $\mathcal{L}$, the starting approximation $V_0(t)$, the auxiliary parameter $\hbar$, and the auxiliary function $H(t)$. The solution $V(t)$ provided by the preceding approach is theoretically dependent on the auxiliary linear operator $\mathcal{L}$, the initial approximation $V_0(t)$, the auxiliary parameter $\hbar$, and the auxiliary function $H(t)$. As a result, unlike all previous analytic procedures, the convergence region and rate of solution series provided by the previously mentioned approach may not be defined uniquely. This is correct, as demonstrated later in this chapter.

4.4.3 Convergence theorem

Theorem 4.4.1. *As long as the series (4.45) converges, where $V_m(t)$ is governed by the high order deformation equations (4.46) and (4.47) under the definitions (4.49) and (4.50), it must be the exact solution of Equations (4.11) and (4.12).*

Proof. If the series

$$\sum_{m=0}^{+\infty} V_m(t)$$

converges, we can write

$$S(t) = \sum_{m=0}^{+\infty} V_m(t)$$

and it holds

$$\lim_{m \to +\infty} V_m(t) = 0 \tag{4.52}$$

Using the definition (4.49) of χ_m, we have

$$\sum_{m=1}^{n} [V_m(t) - \chi_m V_{m-1}(t)]$$
$$= V_1 + (V_2 - V_1) + (V_3 - V_2) + \cdots + (V_n - V_{n-1})$$
$$= N_n(t)$$

which gives us, according to (4.52),

$$\sum_{m=1}^{+\infty} [V_m(t) - \chi_m V_{m-1}(t)] = \lim_{n \to +\infty} V_n(t) = 0.$$

Furthermore, using the above expression and the definition (4.30) of $\mathcal{L}$, we have

$$\sum_{m=1}^{+\infty} \mathcal{L}[V_m(t) - \chi_m V_{m-1}(t)] = \mathcal{L} \lim_{n \to +\infty} [V_m(t) - \chi_m V_{m-1}(t)] = 0.$$

From the above expression and Equation (4.46), we obtain

$$\sum_{m=1}^{+\infty} \mathcal{L}[V_m(t) - \chi_m V_{m-1}(t)] = \hbar\, H(t) \sum_{m=1}^{+\infty} R_m(\vec{V}_{m-1}) = 0.$$

which gives, since $\hbar \neq 0$ and $H(t) \neq 0$, that

$$\sum_{m=1}^{+\infty} R_m(\vec{V}_{m-1}) = 0 \tag{4.53}$$

From (4.50), it holds

$$\sum_{m=1}^{+\infty} R_m(\vec{V}_{m-1}) = \sum_{m=1}^{+\infty} \left[\dot{V}_{m-1}(t) + \sum_{j=1}^{m-1} V_j(t)V_{m-1-j}(t) - (1 - \chi_m) \right] \tag{4.54}$$

$$= \sum_{m=1}^{+\infty} \dot{V}_m(t) - 1 + \sum_{m=1}^{+\infty} \sum_{j=0}^{m-1} V_j(t)V_{m-1-j}(t) \tag{4.55}$$

$$= \sum_{m=1}^{+\infty} \dot{V}_m(t) - 1 + \sum_{j=1}^{+\infty} \sum_{m=j+1}^{\infty} V_j(t)V_{m-1-j}(t) \tag{4.56}$$

$$= \sum_{m=1}^{+\infty} \dot{V}_m(t) - 1 + \sum_{j=1}^{+\infty} V_j(t) \sum_{i=0}^{+\infty} V_i(t) \tag{4.57}$$

$$= \dot{S}(t) + S^2(t) - 1. \tag{4.58}$$

From Equations (4.53) and (4.54), we have

$$\dot{S}(t) + S^2(t) - 1 = 0, \qquad t \geq 0.$$

From Equations (4.29) and (4.47), it holds

$$S(0) = \sum_{m=0}^{+\infty} V_m(0) = V_0(0) + \sum_{m=1}^{+\infty} V_m(0) = V_0(0) = 0.$$

As a result, $S(t)$ must be the exact solution of Equations (4.11) and (4.12) based on the two expressions mentioned above. The proof is now complete.

$\square$

As will be shown later, the aforementioned theorem holds true for the auxiliary linear operator $\mathcal{L}$ described by (4.30) in a somewhat general form, where $\gamma_1(t) \neq 0$ and $\gamma_2(t)$ can be various functions. It's crucial to understand the convergence theorem. This theorem allows us to focus on making sure that the approximation series converges. It is obvious that the auxiliary parameter, the auxiliary function $H(t)$, the initial guess $V_0(t)$, and the auxiliary linear operator $\mathcal{L}$ are all necessary for the series to converge (4.45). Fortunately, the homotopy analysis method gives us a lot of flexibility in choosing any of them. This means that the series (4.45) must converge to the exact solution in the area $0 \leq t \leq t_0$ as long as $\hbar$, $H(t)$, $V_0(t)$, and $\mathcal{L}$ are suitably chosen such that it does so. The foundation of the validity and flexibility of the homotopy analysis approach is therefore established by the convergence theorem in conjunction with the possibility of choice of the auxiliary parameter $\hbar$, the auxiliary function $H(t)$, the initial guess $V_0(t)$, and the auxiliary linear operator $\mathcal{L}$. establishes the cornerstone of the validity and flexibility of the homotopy analysis method.

4.4.4   Some fundamental rules

In order to create the zero-order deformation equation, we have a lot of choice in selecting the auxiliary linear operator $\mathcal{L}$, the initial approximation $V_0(t)$, and the auxiliary function $H(t)$. The preceding freedom is theoretically so great that we are free to select any auxiliary function $H(t)$, any initial approximation $V_0(t)$, and any auxiliary linear operator $\mathcal{L}$. However, from a practical perspective, freedom seems too enormous, and we need a few fundamental rules that will direct us

The goal of analytical approximation is to express a nonlinear problem's solution using the appropriate base functions. It is well-informed that several alternative base functions can approximate a real function $f(x)$, and that a better set of base functions can approximate a real function $f(x)$ more effectively. Therefore, the type of base functions has a significant impact on how effectively a nonlinear problem may be approximated. Choosing a substantially better selection of base functions is essential for effectively approximating a given nonlinear problem. Fortunately, using the freedom in the choice of the auxiliary linear operator $\mathcal{L}$, the initial approximation $V_0(t)$, and the auxiliary function $H(t)$, we can obtain many solution expressions of $V(t)$ presented by different base functions from which we might choose a better one to approximate a given nonlinear problem more efficiently.

Many times, even without addressing a specific nonlinear problem, we may determine what types of base functions are appropriate to describe the answer by analyzing its physical background, initial/boundary conditions, and/or type of nonlinearity. For instance, let

$$\{e_k(t) \mid k = 0, 1, 2, \dots\} \tag{4.59}$$

denote such a collection of base functions for the example issue discussed in this chapter. The solution can be shown as a series.

$$V(t) = \sum_{n=0}^{\infty} c_n \, e_k(t) \tag{4.60}$$

where c_n is the coefficient. The auxiliary function $H(t)$, the initial approximation $V_0(t)$, and the auxiliary linear operator $\mathcal{L}$ must be selected in such a way that all solutions of the corresponding high-order deformation equations exist and can be expressed by this set of base functions once such a set of base functions is established. This gives us a fundamental guideline, the rule of solution expression, to guide the selection of the auxiliary function $H(t)$, the initial approximation $V_0(t)$, and the auxiliary linear operator $\mathcal{L}$. As demonstrated in this chapter, this rule is crucial to the homotopy analysis method's framework.

As previously stated, a real function $f(x)$ can be expressed by a variety of base functions. As a result, multiple types of rule of solution expressions may exist, and any of them may provide valid approximations for a given nonlinear problem. In this scenario, we might be able to get the best one by selecting the best collection of base functions.

To further limit the choice of the auxiliary function $H(t)$, it appears necessary to suggest the *rule of coefficient ergodicity*, which states that all coefficients in the solution expression, such as c_n in (4.60), can be adjusted to assure the completeness of the set of base functions. In many circumstances, auxiliary functions can be uniquely determined using the rule of solution expression and the rule of coefficient ergodicity. This provides the so-called *rule of solution existence.*

rule of solution expression, rule of coefficient ergodicity, and *rule of solution existence* all play key roles in simplifying the application of the homotopy analysis method.

4.4.5   Solution expressions

The solution provided by the homotopy analysis approach, in contrast to the previous solution expressions provided by the perturbation and nonperturbation methods stated above, can be represented by a variety of alternative base functions, as will be seen in this subsection.

Solution expressed by polynomial functions

A power series of t makes up the perturbation solution (4.16). Therefore, using the base functions is simple

$$\{t^{2m+1} \,|\, m = 0, 1, 2, 3, \dots \} \tag{4.61}$$

to represent $V(t)$, i.e.,

$$V(t) = \sum_{m=0}^{+\infty} a_m \, t^{2m+1} \tag{4.62}$$

where a_m is a coefficient. This gives us the first rule for expressing the solution of the example problem.

In accordance with the initial condition (4.29) and the first rule of solution expression, it is simple to choose

$$V_0(t) = t \tag{4.63}$$

as the initial approximation of $V(t)$, and to choose an auxiliary linear operator

$$\mathcal{L}[\Phi(t; q)] = \frac{\partial \Phi(t; q)}{\partial t} \tag{4.64}$$

with the property

$$\mathcal{L}(C_1) = 0, \tag{4.65}$$

where C_1 is a constant for integral. The auxiliary function $H(t)$ can be selected using the first *rule of solution expression* denoted by (4.62) and from Equation (4.46)

$$H(t) = t^{2k}. \tag{4.66}$$

According to (4.65), the solution of Equation (4.46) becomes

$$V_m(t) = \chi_m V_{m-1}(t) = \hbar \int_0^t \tau^{2k} R_m(\vec{V}_{m-1})d\tau + C_1$$

where the initial condition (4.47) determines the integral constant C_1 . The term t^{-1}, which appears when $\kappa \leq -1$, is discovered to violate the first *rule of solution expression*, represented by (4.62) when it appears in the solution expression of $V_m(t)$. Furthermore, when $\kappa \geq 1$, the base t^3 always vanishes from the $V_m(t)$ solution equation, leaving a coefficient for the term t^3, to be zero, that cannot be changed even if the order of approximation increases to infinity. But this defies the coefficient ergodicity criterion. We have to put $\kappa = 0$ in order to comply with both the coefficient ergodicity rule and the first rule of solution expression defined by (4.62). This determines the associated auxiliary function in a unique way as follows

$$H(t) = 1. \tag{4.67}$$

We now successively obtain

$$V_1(t) = \frac{1}{3}\hbar t^3,$$

$$V_2(t) = \frac{1}{3}\hbar(1 + \hbar)t^3 + \frac{2}{15}\hbar^2 t^5,$$

$$V_3(t) = \frac{1}{3}\hbar(1 + \hbar)^2 t^3 + \frac{2}{15}\hbar^2(1 + \hbar)t^5 + \frac{17}{315}\hbar^3 t^7,$$

$$\vdots$$

It is discovered that the corresponding mth-order approximation can be stated by

$$V(t) \approx \sum_{k=0}^{m} V_k(t) = \sum_{n=0}^{m} \mu_0^{m,n}(\hbar)[\alpha_{2n+1}t^{2n+1}], \tag{4.68}$$

where the coefficient α_{2n+1} is the same as that found in the perturbation solution (4.16), and the function $\mu_0^{m,n}(\hbar)$ is defined as

$$\mu_0^{m,n}(\hbar) = (-(\hbar))^n \sum_{j=0}^{m-n} \binom{n-1+j}{j}(1 + \hbar)^j, \tag{4.69}$$

We still have the option to select an appropriate value for the auxiliary parameter $\hbar$ despite the fact that the initial approximation $V_0(t)$, the auxiliary linear operator $\mathcal{L}$, and the auxiliary function $H(t)$ have all been determined. A family of solution expressions for auxiliary parameter $\hbar$ is indicated by equation (4.68). The property of the function $\mu_0^{m,n}(\hbar)$ indicated above can be demonstrated with ease as follows

$$\mu_0^{m,n}(1) = 1, \qquad \text{when } n \leq m. \tag{4.70}$$

For any a finite positive integer n, it holds

$$\lim_{m \to +\infty} \mu_0^{m,n}(\hbar) = \begin{cases} 1 & \text{when} \quad |1 + \hbar| < 1, \\ \infty & \text{when} \quad |1 + \hbar| > 1. \end{cases} \tag{4.71}$$

Later in this chapter, these two properties will be proved. As a result, when $\hbar = -1$, we know from (4.70), (4.68), and (4.16),

$$V(t) = V_{pert}(t). \tag{4.72}$$

As a result, the perturbation solution (4.16) when $\hbar = -1$, the solution (4.19) provided by Lyapunov's artificial small parameter approach, and the solution (4.21) provided by Adomian's decomposition method are all just special cases of the solution expression (4.68). Equation (4.68) is more generic since it logically incorporates the solution expression from the perturbation approach, Lyapunov's artificial small parameter method, and Adomian's decomposition method.

Note that the auxiliary parameter affects how the coefficients of the solution expression (4.68) behave. According to (4.71), $|1 + \hbar| < 1$ is a need for the series (4.68) to be convergent , i.e.,

$$-2 < \hbar < 0.$$

The fact that the convergence zone of the solution series (4.68) depends on the value of $\hbar$ is intriguing. According to Figure (4.2.1), the convergence region of the series (4.68) increases in size the closer the value of $\hbar(-2 < \hbar < 0)$ is near zero. The solution series (4.68), it is discovered, converges in the region.

$$0 \leq t < \rho_0 \sqrt{\frac{2}{|\hbar|} - 1},$$

where $\rho_0 \approx 3/2$ is the convergence radius of the perturbation solution (4.16). So, as $\hbar(-2 < \hbar < 0)$ tends to zero from below, the solution series (4.68) converges to the exact solution $V(t) = \tanh(t)$ in the whole region

$$0 \leq t < +\infty.$$

In contrast to all prior analytical methods, we are able to regulate and alter the convergence region of the solution series (4.68) by putting a suitable value for $\hbar$. Therefore, the auxiliary parameter offers us a convenient means of adjusting and controlling convergence regions of solution series.

Solution expressed by fractional functions

Although the solution expression (4.68) represented by the base functions (4.61) can be valid in the whole region

$$0 \le t < +\infty$$

as $\hbar(-2 < \hbar < 0)$ goes to zero, the order of approximation must be very high to produce an accurate enough answer when the absolute value of $\hbar(-2 < \hbar < 0)$ is small. Although this type of approximation is hardly efficient, Theoretically, it is better and more general than the perturbation method's solution (4.16), Lyapunov's artificial small parameter approach, and Adomian's decomposition method. To approximate $V(t)$ more efficiently, a better set of basis functions must be chosen.

As previously stated, knowing the limit velocity is simple even without solving Equations (4.11) and (4.12)

$$V(+\infty) = 1.$$

This property is obviously not satisfied by the initial approximation (4.63). Because a power series converges in a finite region, the set (4.61) of polynomial functions is not appropriate to efficiently approximate $V(t)$ in the entire region $0 \le t < +\infty$.

Take notice that it holds

$$\lim_{t \to +\infty} \frac{1}{(1 + t)^m} = 0, \qquad m \ge 1.$$

As a result, a function expressed by a set of base functions

$$\{(1 + t)^{-m} | m = 0, 1, 2, 3, \ldots\} \tag{4.73}$$

As $t \to +\infty$, it has a finite value. We can suppose that the solution $V(t)$ can be written down as

$$V(t) = \sum_{m=0}^{+\infty} \frac{b_m}{(1 + t)^m}, \tag{4.74}$$

where b_m is a coefficient to be determined. This provides us with the second *rule of solution expression* of the illustrative problem.

Under the second *rule of solution expression* and using the initial condition (4.12) and the limit velocity (4.13), it is straightforward to choose

$$V_0(t) = 1 - \frac{1}{1+t}, \tag{4.75}$$

as the initial approximation of $V(t)$, and to choose the corresponding auxiliary linear operator

$$\mathcal{L}[\Phi(t; q)] = (1+t)\frac{\partial \Phi(t; q)}{\partial t} + \Phi(t; q) \tag{4.76}$$

with the property

$$\mathcal{L}\left(\frac{C_2}{1+t}\right) = 0, \tag{4.77}$$

where C_2 is an integral constant. Under the definition (4.76) of $\mathcal{L}$, the solution of the high-order deformation equation (4.46) becomes

$$V_m(t) = \chi_m V_{m-1}(t) + \frac{\hbar}{1+t}\int_0^t H(\tau)R_m(\vec{V}_{m-1})d\tau + \frac{C_2}{1+t}, \qquad m \geq 1$$

where the initial condition (4.47) determines the integral constant C_2. According to the second *rule of solution expression* (4.74) and Equation (4.46), the auxiliary function $H(t)$ should be of the type

$$H(t) = \frac{1}{(1+t)^k}, \tag{4.78}$$

where κ is an integer. It is found that, when $\kappa \leq 0$, the solutions of the high-order deformation equation (4.46) contain the term

$$\frac{In(1+t)}{1+t}$$

which, incidentally, violates the second *rule of solution expression*, which is indicated by (4.74). When $\kappa > 1$, the base $(1+t)^{-2}$ vanishes from the $V_m(t)$ solution equation, resulting in a constant coefficient of the term $(1+t)^{-2}$ that cannot be changed even if the order of approximation goes to infinity. However, this defies the so-called coefficient ergodicity criterion. Consequently, we have to select $\kappa = 1$ in order to comply with both the second *rule of solution expression* and *rule of coefficient ergodicity*. This choice determines the corresponding auxiliary function.

$$H(t) = \frac{1}{(1+t)}. \tag{4.79}$$

Thereafter, we successively obtain

$$V_1(t) = -\frac{-\hbar}{1+t} + \frac{-2\hbar}{(1+t)^2} - \frac{-\hbar}{(1+t)^3},$$

$$V_2(t) = -\hbar\left(1 + \frac{7}{12}\hbar\right)\frac{1}{1+t} + \frac{-2\hbar(1+\hbar)}{(1+t)^2}$$

$$- \hbar\left(1 + \frac{7}{2}\hbar\right)\frac{1}{(1+t)^3} + \frac{10\hbar^2}{3(1+t)^4} - \frac{5\hbar^2}{4(1+t)^5}$$

$$\vdots$$

It is found that the corresponding mth-order approximation of $V(t)$ can be expressed by

$$V(t) \approx \sum_{n=0}^{2m+1} \frac{\beta_{m,n}(\hbar)}{(1+t)^n}, \tag{4.80}$$

where $\beta_{m,n}(\hbar)$ is a coefficient dependent upon $\hbar$.

Keep in mind that we are still free to select the auxiliary parameter $\hbar$. In this sense, (4.80) represents a new family of solution expressions. We can first take into account the convergence of certain related series, such as $V'(0), V''(0), V'''(0)$, and so on, in order to evaluate the impact of $\hbar$ on the solution series (4.80). It is discovered that $V'(0) = 1$ holds for each result at all orders of approximations, hence it is impossible for it to offer us any useful information regarding the selection. $V''(0)$ and $V'''(0)$, however, are dependent on $\hbar$. Let $\mathbf{R}_\hbar$ denote a collection of all possible values through which the associated series of $V''(0)$ converges. For the sake of simplicity, we refer to the set $\mathbf{R}_\hbar$ as the valid region of $\hbar$ for $V''(0)$.

4.4.1 states that for every $\hbar \in \mathbf{R}_\hbar$, the corresponding series of $V''(0)$ converges to the same result. A horizontal line segment lies above the valid region $\mathbf{R}_\hbar$ on the curve $V''(0)$ versus $\hbar$. Such a curve, which unmistakably identifies the valid region $\mathbf{R}_\hbar$ of a solution series, is known as the $\hbar$-curve. Based on the solution expression (4.80), the so-called -curves of $V''(0)$ and $V'''(0)$ are as depicted in Figure (4.5). Figure (4.5) makes it obvious that the $V''(0)$ and $V'''(0)$ series provided by the solution series (4.80) are convergent when

$$-3/2 \leq \hbar \leq -1/2.$$

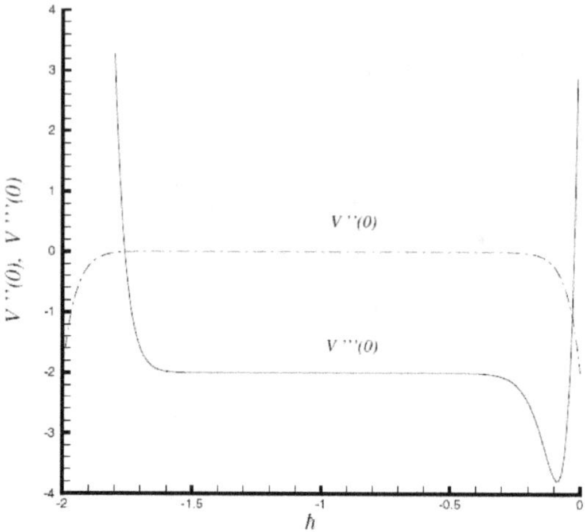

Figure 4.5: The -curve of $V''(0)$ and $V'''(0)$ given by (4.80) when $H(t) = 1/(1+t)$. Dash-dotted line: 20th-order approximation of $V''(0)$; solid line: 20th-order approximation of $V'''(0)$.

This is true. For instance, the series of $V''(0)$ and $V'''(0)$ provided by (4.80) converge to the corresponding exact values 0 and -2, respectively, as shown in Tables (4.4) and (4.5), for five distinct values of $\hbar$ in the region $-3/2 \le \hbar \le -1/2$. It's interesting to note that the value of $\hbar$ has an impact on the approximation series' rate of convergence. According to Tables (4.4) and (4.5), the series of $V''(0)$ and $V'''(0)$ produced by (4.80) converge at the fastest rate when $\hbar = -1$. This suggests that the auxiliary parameter $\hbar$ can be used to modify the convergence rate of the solution series (4.80). It is also true that the series (4.80) converge in the entire region $0 \le t < +\infty$ as long as the series $V''(0)$ and $V'''(0)$ are convergent. Therefore, all of these convergent series must be the exact solution to the original nonlinear problem, as stated by 4.4.1. For instance, the series (4.80, as shown in Table (4.6), converges to the precise solution in the entire region $0 \le t < +\infty$ when $\hbar = -1$. In general, it is simple to determine the associated valid region of $\hbar$ using the so-called $\hbar$-curves. We can guarantee that the associated solution series is convergent by selecting a value of $\hbar$ that is within the valid region. We can regulate and adjust the convergence region and rate of solution series in this way. As a result, the auxiliary parameter $\hbar$ has a significant impact on the homotopy analysis approach.

The solution series (4.80), in contrast to the solution (4.16)), pro-

Table 4.4: Approximations of $V''(0)$ given by (4.80) for different values of $\hbar$.

order	$\hbar = -1/2$	$\hbar = -3/4$	$\hbar = -1$	$\hbar = -5/4$	$\hbar = -3/2$
5	-0.062500	-0.001953	0	-0.001953	0.062500
10	-0.001953	-1.9×10^{-6}	0	-1.9×10^{-6}	-0.001953
15	-0.000061	-1.9×10^{-9}	0	1.9×10^{-9}	0.000061
20	-1.9×10^{-6}	-1.9×10^{-12}	0	-1.9×10^{-12}	-1.9×10^{-6}
25	-6.0×10^{-8}	-1.8×10^{-15}	0	1.8×10^{-15}	6.0×10^{-8}
30	-1.9×10^{-9}	-1.7×10^{-18}	0	-1.7×10^{-18}	-1.9×10^{-9}
35	-5.8×10^{-11}	-1.7×10^{-21}	0	1.7×10^{-21}	5.8×10^{-11}
40	-1.8×10^{-12}	-1.7×10^{-24}	0	-1.7×10^{-24}	-1.9×10^{-12}

Table 4.5: Approximations of $V'''(0)$ given by (4.80) for different values of $\hbar$.

order	$\hbar = -1/2$	$\hbar = -3/4$	$\hbar = -1$	$\hbar = -5/4$	$\hbar = -3/2$
5	-3.312500	-2.138672	-2	-2.251953	-6.937500
10	-2.089844	-2.000278	-2	-1.999516	-1.699219
15	-2.004333	-2.000000	-2	-2.000001	-2.013977
20	-2.000183	-2.000000	-2	-2.000000	-1.99942
25	-2.000007	-2.000000	-2	-2.000000	-2.000023
30	-2.000000	-2.000000	-2	-2.000000	-1.999999
35	-2.000000	-2.000000	-2	-2.000000	-2.000000
40	-2.000000	-2.000000	-2	-2.000000	-2.000000

vided by the perturbation approach, Lyapunov's artificial small parameter method, and Adomian's decomposition method, converges to the precise solution across the entire region $0 \leq t < +\infty$ when $-3/2 \leq \hbar \leq -1/2$. Though theoretically both of them may converge to the precise solution in the entire region $0 \leq t < +\infty$, the solution series (4.80) is thus almost more effective than (4.68). This is primarily due to the fact that base functions (4.73) are better and more effective than (4.61) for the example situation.

Let's look into how the perturbation solution (4.16) and the solution expression (4.80) relate to one another. The substitution of $\hbar = -1$ and

$$\frac{1}{1+t} = 1 - t + t^2 - t^3 + \dots$$

Table 4.6: Comparison of the exact solution (4.14) with the mth-order approximations of $V(t)$ given by (4.80) when $\hbar = -1$.

t	10th-order approx.	20th-order approx.	40th-order approx.	60th-order approx.	exact-result
1/4	0.2449	0.2449	0.2449	0.2449	0.2449
1/2	0.4621	0.4621	0.4621	0.4621	0.4621
3/4	0.6349	0.6351	0.6351	0.6351	0.6351
1	0.7516	0.7616	0.7616	0.7616	0.7616
3/2	0.9082	0.9053	0.9051	0.9051	0.9051
2	0.9720	0.9644	0.9640	0.9640	0.9640
5/2	0.9982	0.9870	0.9866	0.9866	0.9866
3	1.0082	0.9950	0.9950	0.9951	0.9951
4	1.0110	0.9979	0.9992	0.9993	0.9993
5	1.0082	0.9973	0.9997	0.9999	0.9999
10	0.9984	0.9968	1.0003	1.0001	1.0000
100	0.9987	0.9998	1.0001	1.0000	1.0000

in the 10th-order approximation of $V(t)$ given by (4.80), we have

$$V(t) \sim t - \frac{1}{3}t^3 + \frac{2}{5}t^5 - \frac{17}{315}t^7 + \frac{62}{2835}t^9 + \ldots,$$

whose initial few terms match those of the perturbation series (4.16) perfectly.

Solution expressed by exponential functions

It is well known that

$$\lim_{t \to +\infty} \exp(-nt) = 0, \qquad n \geq 1.$$

So, a function expressed by the set of base functions

$$\{\exp(-nt)|\, n \geq 0\} \tag{4.81}$$

as t approaches infinity is finite. The base functions listed above are superior to (4.61) when the limit velocity is taken into account (4.13). If $V(t)$ can be expressed as, then

$$V(t) = \sum_{n=0}^{\infty} c_n \exp(-nt) \tag{4.82}$$

where c_n is a coefficient. This gives us the third *rule of solution expression* for the example problem.

Under the third rule of solution expression and from (4.12) and (4.13), it is simple to choose,

$$V_0(t) = 1 - \exp(-t) \tag{4.83}$$

as the initial approximation of $V(t)$, and to choose the auxiliary linear operator

$$\mathcal{L}[\Phi(t; q)] = \frac{\partial \Phi(t; q)}{\partial t} + \Phi(t; q) \tag{4.84}$$

with the property

$$\mathcal{L}[C_3 \, \text{ext}(-t)] = 0, \tag{4.85}$$

where C_3 is a coefficient of integration. In this scenario, the mth-order deformation equation's (4.46) solution is

$$V_m(t) = \chi_m \, V_{m-1}(t) + \hbar \exp(-t) \int_0^t \exp(\tau) \, H(\tau) \, R_m(\vec{V}_{m-1}) \, d\tau$$
$$+ \, C_3 \exp(-t), \qquad m \geq 1$$

where (4.47) yields a value for the integral constant C_3 . The auxiliary function $H(t)$ should have the following form in accordance with the third *rule of solution expression* represented by (4.82) and from Equation (4.46),

$$H(t) = \exp(-k\,t), \tag{4.86}$$

where κ is an integer. It is discovered that the high-order deformation equations' solutions (4.46) contain the term when $\kappa \leq 0$

$$t \exp(-t),$$

which, incidentally, violates the third *rule of solution expression*, which is indicated by (4.82). The coefficient of the term $\exp(-2t)$ cannot be changed even if the order of approximation increases to infinity as when $\kappa \geq 2$, the base $\exp(-2t)$ always vanishes in the solution expressions of the high-order deformation equation (4.46). However, this violates the alleged coefficient ergodicity criterion. Therefore, in order to comply with the coefficient ergodicity rule as well as the third rule of the solution expression given by (4.82), we had to set $\kappa = 1$, which specifically determines the matching auxiliary function.

$$H(t) = \exp(-t), \tag{4.87}$$

Therefore, we have

$$V_1(t) = -\frac{-\hbar}{2}e^{-t} + \hbar e^{-2t} - \frac{-\hbar}{2}e^{-3t},$$

$$V_2(t) = -\frac{-\hbar}{2}\left(1 + \frac{\hbar}{2}\right)e^{-t} + \hbar\left(1 + \frac{\hbar}{2}\right)e^{-2t} - \frac{-\hbar}{2}(1 + \hbar)e^{-3t}$$
$$\frac{\hbar^2}{2}e^{-4t} - \frac{\hbar^2}{4}e^{-5t},$$

$$\vdots$$

It is found that the corresponding mth-order approximation of $V(t)$ can be generally expressed by

$$V(t) \approx \sum_{n=0}^{2m+1} \gamma_{m,n}(\hbar)\exp(-n\,t) \tag{4.88}$$

where $\gamma_{m,n}(\hbar)$ is a coefficient dependent of $\hbar$

In the auxiliary parameter $\hbar$, equation (4.88) belongs to a family of solution expressions. The so-called $\hbar$-curves of $V''(0)$ and $V'''(0)$ are first plotted, as seen in Figure (4.6), in order to examine the impact of $\hbar$ on the convergence of the solution series (4.88). It is simple to identify the legitimate region of $\hbar$, which corresponds to the line segments almost parallel to the horizontal axis, based on these $\hbar$-curves. As the order of approximations rises, the so-called valid regions of $\hbar$ increase as seen in Figure (4.6). It follows that if a sequence of $V''(0)$ and $V'''(0)$ given by (4.88) belongs to the corresponding valid regions of $\hbar$, then it converges. They must converge to the exact values of $V''(0)$ and $V'''(0)$, respectively, according to Theorem 4.4.1. As seen in Tables (4.7) and (4.8) when $\hbar = -3/2, -5/4, -1, -3/4$ and $-1/2$, this is in fact the case. Take note that the series appears to converge fast at $\hbar = -1$. Additionally, across the entire area $0 \le t < +\infty$, the corresponding solution series of $V(t)$ (4.88) similarly converges to the precise solution (4.14) as long as the series of $V''(0)$ and $V'''(0)$ converge. For example, the approximation result of $V(t)$ given by (4.88), when $\hbar = -1$, agrees well with the exact result (4.14), as shown in Table (4.9). Generally, it is convenient to investigate the influence of $\hbar$ on the convergence of solution series by means of such kinds of $\hbar$-curves.

In the region $0 \le t < +\infty$, it should be noted that the solution equation (4.88) holds true. With the same value of $\hbar$, the solution series (4.88) converges faster than (4.80, and even the 10th-order approximation of (4.88) when $\hbar = -1$ agrees well with the precise solution. This

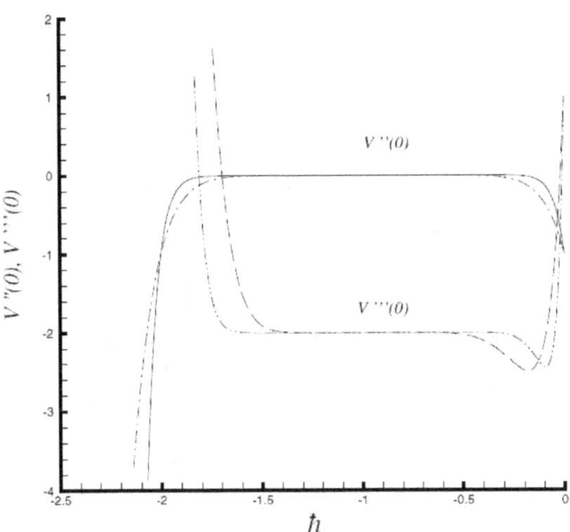

Figure 4.6: The $\hbar$-curves of $V''(0)$ and $V'''(0)$ given by (4.88) when $H(t) = \exp(-t)$. Dash-dotted line: 10th-order approximation of $V''(0)$; solid line: 20th-order approximation of $V''(0)$; dashed lined: 10th-order approximation of $V'''(0)$; dash-dot-dotted line: 20th-order approximation of $V'''(0)$.

is demonstrated by comparing Tables (4.7) to (4.9) with Tables (4.4) to (4.6), respectively. As a result, (4.88) is more effective than (4.80 and, as was already mentioned, (4.80 is more effective than (4.68). These demonstrate how we can more effectively approximate a given nonlinear problem within the framework of the homotopy analysis method by selecting an improved set of base functions.

It is found that the explicit expression of the mth-order approximation (4.88) of $V(t)$ is

$$V(t) \approx 1 + 2 \sum_{n=1}^{m} [(-1)^n \exp(-n\,t)] \mu_0^{m,n} \left(\frac{\hbar}{2}\right) \\ - \exp(-t) \left[\left(1 + \frac{\hbar}{2}\right) + \frac{\hbar}{2} \exp(-2t)\right]^m, \tag{4.89}$$

where (4.69) defines the function $\mu_0^{m,n}(x)$. The recurrence of the function $\mu_0^{m,n}(\hbar)$ is intriguing. Given the property (4.70), the prior equation becomes true when $\hbar = -2$.

$$V(t) \approx 1 + 2 \sum_{n=1}^{m} (-1)^n \exp(-nt) + (-1)^{m+1} \exp[-(2m+1)t]. \tag{4.90}$$

Table 4.7: Comparison of the exact solution (4.14) with the mth-order approximations of $V(t)$ given by (4.80) when $\hbar = -1$.

order	$\hbar = -1/2$	$\hbar = -3/4$	$\hbar = -1$	$\hbar = -5/4$	$\hbar = -3/2$
5	-0.031250	-0.000977	0	-0.000977	0.031250
10	-0.000977	-9.5×10^{-7}	0	-9.5×10^{-7}	-0.000977
15	-0.000031	-9.3×10^{-10}	0	9.3×10^{-10}	0.000031
20	-9.5×10^{-7}	-9.1×10^{-13}	0	-9.1×10^{-13}	-9.5×10^{-7}
25	-3.0×10^{-8}	-8.9×10^{-16}	0	8.9×10^{-16}	3.0×10^{-8}
30	-9.3×10^{-10}	-8.7×10^{-19}	0	-8.7×10^{-19}	-9.3×10^{-10}
35	-2.9×10^{-11}	-8.5×10^{-22}	0	8.5×10^{-22}	2.9×10^{-11}
40	-9.1×10^{-13}	-8.3×10^{-25}	0	-8.3×10^{-25}	-9.1×10^{-13}

Table 4.8: Approximations of $V'''(0)$ given by (4.88) for different values of $\hbar$.

order	$\hbar = -1/2$	$\hbar = -3/4$	$\hbar = -1$	$\hbar = -5/4$	$\hbar = -3/2$
5	-2.375000	-2.041016	-2	-2.076172	-3.500000
10	-2.026367	-2.000083	-2	-1.999854	-1.909180
15	-2.001282	-2.000000	-2	-2.000001	-2.004211
20	-2.000054	-2.000000	-2	-2.000000	-1.999825
25	-2.000002	-2.000000	-2	-2.000000	-2.000007
30	-2.000000	-2.000000	-2	-2.000000	-2.000000
35	-2.000000	-2.000000	-2	-2.000000	-2.000000
40	-2.000000	-2.000000	-2	-2.000000	-2.000000

Note that the exact solution (4.14) can be expanded as a series

$$V(t) \approx 1 + 2 \sum_{n=1}^{+\infty}(-1)^n\exp(-nt), \qquad (4.91)$$

Where, in the region $0 < t < +\infty$, it converges to the exact solution, but diverges at the point $t = 0$ where it either yields 1 or -1. But using an additional term

$$(-1)^{m+1}\exp[-(2m+1)]$$

In the entire region $0 \leq t < +\infty$ including the point $t = 0$, the expression (4.90) converges to the exact answer. In fact, even the $V(t)$ third-order approximate provided by (4.90)), i.e.,

$$V(t) \approx 1 - 2\exp(-2t) + 2\exp(-4t) - 2\exp(-6t) + \exp(-7t), \quad (4.92)$$

Table 4.9: Comparison of the exact solution (4.14) with the mth-order approximations of $V(t)$ given by (4.88) when $\hbar = -1$.

t	10th-order approx.	20th-order approx.	40th-order approx.	60th-order approx.	exact result
1/4	0.2449	0.2449	0.2449	0.2449	0.2449
1/2	0.4619	0.4621	0.4621	0.4621	0.4621
3/4	0.6342	0.6351	0.6351	0.6351	0.6351
1	0.7596	0.7616	0.7616	0.7616	0.7616
3/2	0.9020	0.9051	0.9051	0.9051	0.9051
2	0.9612	0.9639	0.9640	0.9640	0.9640
5/2	0.9845	0.9866	0.9866	0.9866	0.9866
3	0.9937	0.9950	0.9951	0.9951	0.9951
4	0.9988	0.9993	0.9993	0.9993	0.9993
5	0.9997	0.9999	0.9999	0.9999	0.9999
10	1.0000	1.0000	1.0000	1.0000	1.0000
100	1.0000	1.0000	1.0000	1.0000	1.0000

agrees very well with the exact solution, as shown in Figure (4.7).

The idea to avoid the appearance of the term such as

$$ln(1+t)/(1+t), \qquad t\exp(-t)$$

in approximate expansions is not new. In order to obtain uniformly valid approximations, some perturbation techniques were devised to prevent the appearance of the so-called secular terms, such as

$$t\sin t, \; t\cos t$$

Many scientists from the 19th century used this type of technique, including Lindstedt, Bohlin, Poincar'e, Gyld'en, and others. Numerous scientists, including Lighthill, Malkin, Kuo , and Tsien, went on to further develop the concept. The expressions $\ln(1+t)/(1+t)$ and $t\exp(-t)$ nonetheless tend to zero as $t \to +\infty$. These concepts do not, therefore, fall under the umbrella of the secular term used in perturbation techniques. Therefore, it is possible to think of the *rule of solution expression* as a generalization of this concept.

We highlight the fact that $V(t))$ may be expressed by the base function to demonstrate that the term $t\exp(-t)$ in fact does not belong to the so-called secular terms in perturbation methods.

$$\{t^m \exp(-nt) | m \geq 0, n \geq 1\} \tag{4.93}$$

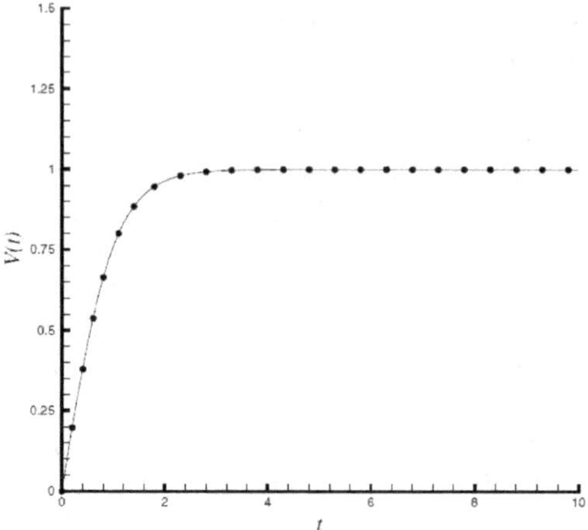

Figure 4.7: The comparison of the third-order approximation (4.92) of $V(t)$ with the exact solution (4.14). Solid line: third-order approximation (4.92); symbols: exact solution (4.14).

The initial approximation is the same as in (4.83), the auxiliary linear operator is the same as in (4.84), but the auxiliary function.

$$H(t) = 1 \qquad (4.94)$$

We can obtain the matching mth-order approximation of $V(t)$ in a manner similar to (4.87), which is formally represented by

$$V(t) \approx 1 + 2 \sum_{n=1}^{m+1} \sum_{k=0}^{m+1-n} \sigma_0^{m,n,k}(\hbar) \left[(-1)^n \frac{(-nt)^k}{k!} \exp(-nt) \right], \qquad (4.95)$$

where

$$\sigma_0^{m,n,k}(\hbar) = \frac{1}{2}[\mu_0^{m,n+k}(\hbar) + \mu_0^{m,n+k-1}(\hbar)]. \qquad (4.96)$$

It is interesting that the function $\mu_0^{m,n}(\hbar)$ appears once again. By means of the so-called $\hbar$-curves of the corresponding $V''(0)$ and $V'''(0)$, it is found that, when $-2 < \hbar < 0$, the solution series (4.95) throughout the whole region $0 \le t < +\infty$ converges to the exact solution (4.14), as shown in Table (4.10)

It should be noted that although the illustrative problem only has a unique solution, the answer $V(t)$ may be described by four different base functions (4.61), (4.73), (4.81), and (4.93) within the context of the

Table 4.10: Comparison of the exact solution (4.14) with the mth-order approximations of $V(t)$ given by (4.95) when $\hbar = -1$.

t	10th-order approx.	20th-order approx.	40th-order approx.	60th-order approx.	exact result
1/4	0.2449	0.2449	0.2449	0.2449	0.2449
1/2	0.4621	0.4621	0.4621	0.4621	0.4621
3/4	0.6351	0.6351	0.6351	0.6351	0.6351
1	0.7616	0.7616	0.7616	0.7616	0.7616
3/2	0.9051	0.9051	0.9051	0.9051	0.9051
2	0.9640	0.9640	0.9640	0.9640	0.9640
5/2	0.9866	0.9866	0.9866	0.9866	0.9866
3	0.9953	0.9950	0.9951	0.9951	0.9951
4	0.9990	0.9993	0.9993	0.9993	0.9993
5	0.9975	0.9999	0.9999	0.9999	0.9999
10	1.0021	0.9982	0.9999	1.0000	1.0000
100	1.0000	1.0000	1.0000	1.0000	1.0000

homotopy analysis method. We get four families of solution expressions as a result, including (4.68), (4.80), (4.4.5), and (4.95). Theoretically, across the entire region $0 \le t < +\infty$, they can all converge to the same exact solution $V(t) = \tanh(t)$

However, the solution expression (4.68), which converges in a finite region for a specific value of $-2 < \hbar < 0$ is the worst since it is the least effective of the four solution expressions. The solution expression (4.4.5) based on the pure exponential functions is more effective than the solution expressions (4.80) based on fractional functions and the solution expression (4.95) based on combined polynomial and exponential functions, and is therefore the best, when Tables(4.6), 4.9, and 4.10 are compared to one another. In comparison to the solution expression (4.80), the solution expression (4.95) is more effective.

The solution to a particular nonlinear problem can be stated by a variety of base functions in the context of the homotopy analysis method, and even if the solution is unique, a better selection of base functions can more effectively approximate it.

In fact, the exact solution to this illustrative example is known, and it is fairly straightforward. It does, however, demonstrate how the homotopy analysis method allows for the adjustment and control of the convergence region and rate of solution series by plotting the so-called $\hbar$-curves and then selecting $\hbar$ in the appropriate valid regions of $\hbar$. Even though expanding convergence regions is unnecessary, by choosing the right value $\hbar$, we can provide a solution series that is more effective. The rule of solution expression and the rule of coefficient ergodicity play significant roles in determining the initial approximation, the auxiliary linear operator, and the auxiliary function, as shown by the illustrative example.

4.4.6 The role of the auxiliary parameter $\hbar$

As previously indicated, the homotopy analysis method is based on the homotopy, a basic topological concept. The so-called zero-order deformation equation, which provides a more general homotopy than the conventional one, is built using the nonzero auxiliary parameter $\hbar$. Consequently, unlike all previous analytical methods, the homotopy analysis method gives us a family formulas for solutions in the auxiliary parameter $\hbar$. The convergence region and rate of solution series are consequently reliant on the auxiliary parameter $\hbar$ and can be significantly increased by selecting an appropriate value for $\hbar$. This gives us an easy approach to modify and regulate the convergence region and rate of the solution series produced by the homotopy analysis method, as shown above.

In this section, we demonstrate that series convergence regions can really be adjusted and controlled by the addition of an auxiliary parameter. This demonstration can give us a logical justification for the validity of the homotopy analysis technique.

First, we want to emphasise that the definition (4.69) of $\mu_0^{m,n}$ in the solution formulations (4.68), (4.4.5), and (4.95) is derived from the homotopy analysis approach. It's interesting that the well-known Newtonian binomial theorem directly leads to the same definition. Consider a series to demonstrate this,

$$\frac{1}{1+t} = 1 - t + t^2 - t^3 + \cdots = \lim_{m \to +\infty} \sum_{n=0}^{m} (-1)^n t^n, \ |t| < 1. \qquad (4.97)$$

Define

$$x = 1 + \hbar + \hbar t,$$

which gives

$$\frac{1}{1+t} = -\frac{\hbar}{(1-x)}$$

When $|x| = |1 + \hbar + \hbar t| < 1$ and $|1 + \hbar| < 1$, i.e.,

$$-1 < t < \frac{2}{\hbar} - 1, \qquad -2 < \hbar < 0,$$

it holds

$$\frac{1}{1+t} = -\frac{\hbar}{(1-x)} = -\hbar(1 + x + x^2 + x^3 + \dots) = -\hbar \sum_{n=0}^{+\infty}(1 + \hbar + \hbar\,t)^n.$$

Thus,

$$\frac{1}{1+t} = \lim_{m\to+\infty}\left[-\hbar\sum_{n=0}^{m}(1 + \hbar + \hbar\,t)^n\right]$$

is valid in the region

$$-1 < t < \frac{2}{|\hbar|} - 1, \qquad -2 < \hbar < 0,$$

We have

$$-\hbar\sum_{n=0}^{m}(1 + \hbar + \hbar\,t)^n$$

$$= \sum_{n=0}^{m}\sum_{k=0}^{n}\binom{n}{k}(1+\hbar)^{n-k}(\hbar\,t)^k$$

$$= \sum_{k=0}^{m}\sum_{n=k}^{m}\binom{n}{k}(1+\hbar)^{n-k}\hbar^k\,t^k$$

$$= \sum_{k=0}^{m}(-1)^k\,t^k\,(-\hbar)^{k+1}\sum_{i=0}^{m-k}\binom{k+i}{k}(1+\hbar)^i$$

$$= \sum_{k=0}^{m}(-1)^k\,t^k\left[(-\hbar)^{k+1}\sum_{i=0}^{m-k}\binom{k+i}{k}(1+\hbar)^i\right]$$

$$= \sum_{n=0}^{m}(-1)^n\,t^n\,\mu_{-1}^{m,n}(\hbar),$$

where

$$\mu_{-1}^{m,n}(\hbar) = (-\hbar)^{n+1}\sum_{j=0}^{m-n}\binom{n+j}{j}(1+\hbar)^j. \tag{4.98}$$

Comparing the above to the definition (4.69), we gain the relationship

$$\mu_{-1}^{m,n}(\hbar) = \mu_{-1}^{m+1,n+1}(\hbar). \tag{4.99}$$

Thus, it holds

$$\frac{1}{1+t} = \lim_{m\to+\infty}\sum_{n=0}^{m}\mu_{0}^{m+1,n+1}(\hbar)\left[(-1)^n\,t^n\right] \tag{4.100}$$

in the region

$$-1 < t < \frac{2}{|\hbar|} - 1, \qquad (-2 < \hbar < 0),$$

It is clear that the convergence region is equal to $-1 < t < 1$ when $\hbar = -1, -1 < t < 3$ when $\hbar = -1/2$, and $-1 < t < 99$ when $\hbar = -1/50$. The convergence region in particular becomes

$$11 < t < +\infty$$

as $\hbar$ tends to zero from below. As a result, the auxiliary parameter $\hbar$ can be used to modify and regulate the series' convergence region (4.100).

We must point out that in this case, the same definition, $\mu_0^{m,n}(\hbar)$, is first discovered within the framework of the homotopy analysis method and is then inferred from the renowned Newtonian binomial theorem. The validity and reasonability of the homotopy analysis method are rationally demonstrated by this fact.

The aforementioned concepts can be used to provide a generalised theorem like the one that follows.

Theorem 4.4.2. *It holds*

$$(1 + t)^\alpha = \lim_{m \to +\infty} \sum_{n=0}^{m} \mu_\alpha^{m,n}(\hbar) \begin{pmatrix} \alpha \\ n \end{pmatrix} t^n \qquad (4.101)$$

for a real number $\alpha(\alpha \neq 0, 1, 2, 3, \dots)$ in the region

$$-1 < t < \frac{2}{|\hbar|} - 1, \qquad (-2 < \hbar < 0),$$

where

$$\begin{pmatrix} \alpha \\ n \end{pmatrix} = \frac{\alpha(\alpha - 1)(\alpha - 2)\dots(\alpha - n + 1)}{n!}$$

and

$$\mu_\alpha^{m,n}(\hbar) = (\hbar)^{n-\alpha} \sum_{j=0}^{m-n} (-1)^j \begin{pmatrix} \alpha - n \\ j \end{pmatrix} (1 + \hbar)^j. \qquad (4.102)$$

Proof. Write $x = 1 + \hbar + \hbar t$. Let $|x| < 1$ and $|1 + \hbar| < 1$, i.e.,

$$-1 < t < \frac{2}{|\hbar|} - 1, \qquad (-2 < \hbar < 0),$$

By the traditional Newton binomial theorem, it holds when $|x| < 1$ and $|1 + \hbar| < 1$ that

$$(1 + t)^\alpha = (-\hbar)^{-\alpha}(1 - x)^\alpha = (-\hbar^{-\alpha}) \sum_{n=0}^{+\infty} (-1)^n \binom{\alpha}{n} x^n$$

$$= (-\hbar^{-\alpha}) \sum_{n=0}^{+\infty} (-1)^n \binom{\alpha}{n} (1 + \hbar + \hbar t)^n$$

$$= \lim_{m \to +\infty} (-\hbar^{-\alpha}) \sum_{n=0}^{m} (-1)^n \binom{\alpha}{n} (1 + \hbar + \hbar t)^n.$$

The sum of the first m terms of above series is given by

$$(-\hbar^{-\alpha}) \sum_{n=0}^{m} (-1)^n \binom{\alpha}{n} (1 + \hbar + \hbar t)^n$$

$$= (-\hbar^{-\alpha}) \sum_{n=0}^{m} (-1)^n \binom{\alpha}{n} \sum_{j=0}^{n} \binom{n}{j} (1 + \hbar)^{n-j} \hbar^j t^j$$

$$= (-\hbar^{-\alpha}) \sum_{j=0}^{n} t^j \sum_{n=j}^{m} (-1)^n \binom{\alpha}{n} \binom{n}{j} (1 + \hbar)^{n-j} \hbar^j$$

$$= (-\hbar^{-\alpha}) \sum_{j=0}^{m} t^j \sum_{i=0}^{m-j} (-1)^{i+j} \binom{\alpha}{i+j} \binom{i+j}{j} (1 + \hbar)^i \hbar^j$$

$$= (-\hbar^{-\alpha}) \sum_{j=0}^{m} t^j \sum_{i=0}^{m-j} (-1)^{i+j} \binom{\alpha}{j} \binom{\alpha-j}{i} (1 + \hbar)^i \hbar^j$$

$$= \sum_{j=0}^{m} \left[\binom{\alpha}{j} t^j \right] \sum_{i=0}^{m-j} (-1)^i \binom{\alpha-j}{i} (1 + \hbar)^i (-\hbar)^{j-\alpha}$$

$$= \sum_{n=0}^{m} \mu_\alpha^{m,n}(\hbar) \left[\binom{\alpha}{n} t^n \right],$$

where

$$\mu_\alpha^{m,n}(\hbar) = (-\hbar)^{n-\alpha} \sum_{j=0}^{m-n} (-1)^j \binom{\alpha-n}{j} (1 + \hbar)^j.$$

This ends the proof. $\qquad\qquad\square$

The definition (4.102) is valid for all real numbers even though it can only be deduced for real numbers $-\infty < \alpha < +\infty$ with the exception of integers $\alpha = 0, 1, 2, 3, \ldots,$ Using the definition (4.102), we obtain that for any integer k.

$$\mu_k^{m,n}(\hbar) = (-\hbar)^{n-k} \sum_{j=0}^{m-n} \binom{n-k-1+j}{j} (1+\hbar)^j, \qquad (4.103)$$

which includes the definitions (4.69) of $\mu_0^{m,n}(\hbar)$ and the definition (4.98) of $\mu_{-1}^{m,n}(\hbar)$. It can be proven that it holds for any real number $\alpha \in (-\infty, +\infty)$

$$\mu_\alpha^{m,n}(-1) = 1 \qquad (4.104)$$

and

$$\lim_{m \to +\infty} \mu_\alpha^{m,n}(\hbar) = 1, \qquad \text{when } |1+\hbar| < 1 \qquad (4.105)$$

for every positive finite integer n. The definitions (4.69) and (4.102) state that the statement is true for the number $l \geq 0$.

$$\mu_{-1}^{m,n}(\hbar) = \mu_0^{m+l,n+l}(\hbar) \qquad (4.106)$$

It is straightforward to prove 4.104. When $|1+\hbar| < 1$, the definition (4.102) holds that

$$\lim_{m \to +\infty} \mu_\alpha^{m,n}(\hbar)$$

$$= (-\hbar)^{n-\alpha} \sum_{k=0}^{+\infty} (-1)^k \binom{\alpha-n}{k} (1+\hbar)^k$$

$$= (-\hbar)^{n-\alpha} \sum_{k=0}^{+\infty} \binom{\alpha-n}{k} (-1-\hbar)^k$$

$$= (-\hbar)^{n-\alpha} [1 + (-1-\hbar)]^{\alpha-n}$$

$$= 1$$

This ends the proof of (4.105).

It is important to note that when $\alpha = 0$ and $\alpha = -1$, respectively, the expressions for $\mu_0^{m,n}(\hbar)$ defined by (4.69) and $\mu_{-1}^{m,n}(\hbar)$ defined by (4.98) are just special instances of (4.102). The validity and reasonability of the answers (4.68), (4.4.5) and (4.95) provided by the homotopy analysis approach are logically confirmed by all of them.

Another explanation is possible for the logic of the solutions (4.95). Any given function's Taylor series is known to be unique. The property (4.71) states that when $|1+\hbar| < 1$, for any given finite positive number N, it holds,

$$\lim_{m \to +\infty} \sum_{n=0}^{N} (\alpha_{2n+1} t^{2n+1}) \mu_0^{m,n}(\hbar) = \sum_{n=0}^{N} \alpha_{2n+1} t^{2n+1}.$$

Therefore, if the order of approximation increases to infinity for any finite positive integer N, the sum of the first N terms of the solution (4.68) is equal to the perturbation solution sums of the first N terms (4.16). Thus, the series (4.68) complies with the Taylor series' uniqueness, albeit in a more general meaning. Let

$$(\alpha_1, \alpha_3, \alpha_5, \alpha_7, \dots)$$

indicate a point in the space $\mathcal{S}$ where $\alpha_k (k = 1, 3, 5, \dots)$ is the coefficient in the perturbation solution (4.16). The perturbation solution (4.16), on the other hand, can be viewed as an approach to the point $(\alpha_1, \alpha_3, \alpha_5, \alpha_7, \dots)$, along a conventional path Γ_0. Denote a point in a space $\mathcal{S}$, where $\alpha_k (k = 1, 3, 5, \dots)$ is the coefficient in the perturbation solution (4.16). The perturbation solution (4.16) can be regarded as an approach to the point $(\alpha_1, \alpha_3, \alpha_5, \alpha_7, \dots)$ along such a traditional path Γ_0 defined by:

$$(\alpha_1, 0, 0, 0, \dots),$$
$$(\alpha_1, \alpha_3, 0, 0, \dots),$$
$$(\alpha_1, \alpha_3, \alpha_5, 0, \dots),$$
$$\vdots$$

However, the solution series (4.68) can be regarded as an approach to the same point

$$(\alpha_1, \alpha_3, \alpha_5, \alpha_7, \dots)$$

but along such a more general path $\Gamma(\hbar)$ defined by:

$$(\alpha_1 \mu_0^{0,0}, 0, 0, 0, \dots),$$
$$(\alpha_1 \mu_0^{1,0}, \alpha_3 \mu_0^{1,1}, 0, 0, \dots),$$
$$(\alpha_1 \mu_0^{2,0}, \alpha_3 \mu_0^{2,1}, \alpha_5 \mu_0^{2,2}, 0, \dots),$$
$$\vdots$$

It is important to note that the path $\Gamma(\hbar)$ is dependent on the auxiliary parameter $\hbar$. The path $\Gamma(-1)$ (when $\hbar = -1$) is identical to the conventional one Γ_0, according to (4.70). The path $\Gamma(\hbar)$ deviates from the standard path Γ_0 when $|1 + \hbar| < 1$ but $\hbar \neq -1$. Even in this situation, they all approach the same point $(\alpha_1, \alpha_3, \alpha_5, \alpha_7, \dots)$, in accordance with property (4.71). Since there are an unlimited number of possible approaches to the same point $(\alpha_1, \alpha_3, \alpha_5, \alpha_7, \dots)$, the solution series (4.68) can be seen as a type of limit process. It is common knowledge that the outcome of this type of limit process frequently depends on the path that is being followed. For instance, consider the limit.

$$\lim_{(x,y)\to(0,0)} \frac{\sqrt{x^2 + y^2}}{|x|}$$

There are numerous paths to get to $(0,0)$. Let's consider the path $y = \beta x$ for the sake of simplicity, where β is a real value. Obviously, it holds that

$$\lim_{(x,y)\to(0,0)} \frac{\sqrt{x^2 + y^2}}{|x|} = \sqrt{1 + \beta^2}.$$

Therefore, the limit depends on the incoming path to point $(0,0)$. Because the function $\mu_0^{m,n}(\hbar)$ creates different approaching paths by different values of $\hbar$, it is evident why the convergence zone of the solution series (4.68) depends on the auxiliary parameter $\hbar$.

The function $\mu_\alpha^{m,n}(\hbar)$ can be used to define different approaching pathways for a limit process by changing values of α and $\hbar$, as explained above. The function $\mu_\alpha^{m,n}(\hbar)$ is referred to as the *approach function of the first kind* for the sake of this reasoning. In order to broaden the definition of $\sigma_0^{m,n}(\hbar)$ (4.96), we define

$$\sigma_\alpha^{m,n,k}(\hbar) = \frac{1}{2}[\mu_\alpha^{m,n+k}(\hbar) + \mu_\alpha^{m,n+k-1}(\hbar)] \tag{4.107}$$

as *the approach function of the second kind*, where $|1+\hbar| < 1$ and $-\infty < \alpha < +\infty$. It is easy to prove that, for $\alpha \in (-\infty, +\infty)$ and $0 \leq n \leq m+1$, it holds

$$\sigma_\alpha^{m,n,k}(-1) = \begin{cases} 1, & \text{when} \quad 0 \leq k < m+1-n, \\ 1/2, & \text{when} \quad k = m+1-n. \end{cases} \tag{4.108}$$

and

$$\lim_{m\to+\infty} \sigma_\alpha^{m,n,k}(\hbar) = \begin{cases} 1, & \text{when} \quad |1+\hbar| < 1, \\ \infty, & \text{when} \quad |1+\hbar| > 1, \end{cases} \tag{4.109}$$

where n and k are finite positive integers.

Since the approach functions $\mu_\alpha^{m,n}(\hbar)$ and $\sigma_\alpha^{m,n,k}(\hbar)$ have relatively general definitions, they can be used to significantly enlarge the convergence regions of approximation series. As an illustration, consider the traditional Taylor series.

$$\sum_{n=0}^{+\infty} \frac{f^{(n)}(z_0)}{n!}(z-z_0)^n$$

of a function $f(z)$, we can define the *generalized Taylor series* of the first kind

$$\lim_{m\to+\infty} \sum_{n=0}^{m} \mu_\alpha^{m,n}(\hbar)\left[\frac{f^{(n)}(z_0)}{n!}(z-z_0)^n\right]$$

and *the generalized Taylor series* of the second kind

$$\lim_{m\to+\infty} \sum_{n=0}^{m} \sigma_\alpha^{m,n,0}(\hbar)\left[\frac{f^{(n)}(z_0)}{n!}(z-z_0)^n\right]$$

where (4.102) and (4.107) define $\mu_\alpha^{m,n}(\hbar)$ and $\sigma_\alpha^{m,n,0}$ respectively. By selecting appropriate values for $\hbar$ and α, the convergence zones of these generalized Taylor series could be significantly expanded. Using the mth-order approximations (4.68), (4.4.5), and (4.95) as an example, we can generalize

$$V(t) \approx \sum_{n=0}^{m} \mu_\alpha^{m,n}(\hbar)[\alpha_{2n+1}t^{2n+1}], \qquad (4.110)$$

$$V(t) \approx 1 + 2\sum_{n=1}^{m}[(-1)^n\exp(-nt)]\mu_\alpha^{m,n}\left(\frac{\hbar}{2}\right)$$
$$- \exp(-t)\left[\left(1+\frac{\hbar}{2}\right) + \frac{\hbar}{2}\exp(-2t)\right]^m, \quad (4.111)$$

and

$$V(t) \approx 1 + 2\sum_{n=1}^{m+1}\sum_{k=0}^{m+1-n} \sigma_\alpha^{m,n,k}(\hbar)\left[(-1)^n\frac{(-nt)^k}{k!}\exp(-nt)\right], \quad (4.112)$$

respectively, where $|1+\hbar| < 1$ and $\alpha \in (-\infty, +\infty)$.

For instance, when $\alpha = \pi/4$, as shown in Figure (4.8), the convergence region of the approximation (4.110 as $\hbar$ tends to zero from below and grows increasingly wider. Additionally, as shown in Table (4.11), all of the 20th-order estimates of (4.111) provided by $\alpha = \pm 1/2, \pm \pi/4$ coincide with the correct solution when $\hbar = -1$. When $\hbar = -1/2$, as shown in Table (4.12), all of the 20th-order approximations of (4.112) provided by $\alpha = \pm 1/2, \pm \pi/4$ agree. Therefore, the meaning of the functions $\mu_\alpha^{m,n}(\hbar)$ and $\sigma_\alpha^{m,n,k}(\hbar)$ is really quite general.

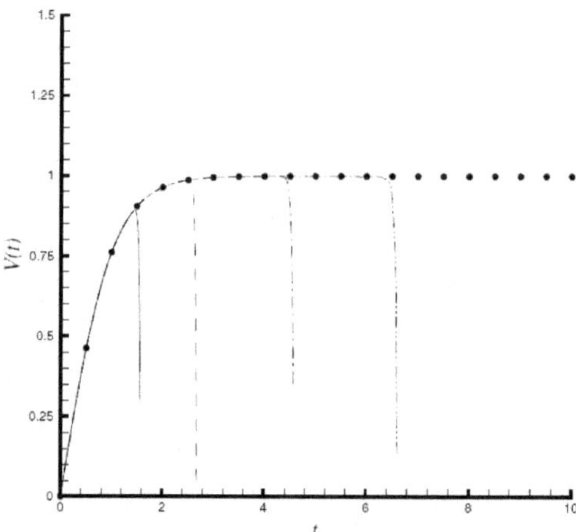

Figure 4.8: Comparison of the exact solution (4.14) with the solution expression (4.110) at the 31st-order of approximation when $\alpha = \pi/4$. Symbols: exact solution; solid line: (4.110) when $\hbar = -1$; dashed line: (4.110) when $\hbar = -1/2$; dash-dotted line: (4.110) when $\hbar = -1/5$; dash-dot-dotted line: (4.110) when $\hbar = -1/10$.

In this part, we emphasize how the Newtonian binomial theorem can be used to independently derive the definition (4.69) of $\mu_0^{m,n}(\hbar)$ (which was first found in the homotopy analysis technique). Furthermore, we demonstrate that adding an auxiliary parameter does really allow for adjustment and control of the region and rate of a series convergence. We also draw attention to the fact that different approaching pathways are defined by different values of $\hbar$ and α and in the functions $\mu_\alpha^{m,n}(\hbar)$ and $\sigma_\alpha^{m,n,k}(\hbar)$. These all give us a logical foundation for the applicability of the homotopy analysis method.

Table 4.11: Comparison of the exact solution (4.14) with the 20th-order approximations of $V(t)$ given by (4.111) when $\hbar = -1$ and $\alpha = \pm 1/2, \pm \pi/4$.

t	when $\alpha = -\pi/4$	when $\alpha = -1/2$	when $\alpha = 1/2$.	when $\alpha = \pi/4$.	exact-result
1/4	0.2449	0.2449	0.2449	0.2449	0.2449
1/2	0.4621	0.4621	0.4621	0.4621	0.4621
3/4	0.6351	0.6351	0.6351	0.6351	0.6351
1	0.7616	0.7616	0.7616	0.7616	0.7616
3/2	0.9051	0.9051	0.9051	0.9051	0.9051
2	0.9640	0.9640	0.9640	0.9640	0.9640
5/2	0.9866	0.9866	0.9866	0.9866	0.9866
3	0.9951	0.9951	0.9951	0.9951	0.9951
4	0.9993	0.9993	0.9993	0.9993	0.9993
5	0.9999	0.9999	0.9999	0.9999	0.9999
10	1.0000	1.0000	0.9999	1.0000	1.0000
100	1.0000	1.0000	1.0000	1.0000	1.0000

4.4.7   Homotopy-Padé method

The homotopy analysis method is based on the idea that for the example case, the series (4.43) of $\Phi(t; q)$ converges at $q = 1$. Fortunately, as was already said, we have a lot of freedom to select the initial approximation $V_0(t)$, the auxiliary linear operator $\mathcal{L}L$, the auxiliary function $H(t)$, and the auxiliary parameter $\hbar$ in the context of the homotopy analysis approach. As was previously illustrated, the series (4.43) can converge at $q = 1$ if they are all wisely chosen. Additionally, the auxiliary parameter affects the convergence region and rate of the solution series determined by the homotopy analysis approach. Thus, as illustrated above, the auxiliary parameter offers us a practical means of adjusting and controlling the convergence region and rate of solution series.

There are methods to quicken the convergence of a certain series. The so-called Padé method is one of them that is frequently used. For any given series,

$$\sum_{n=0}^{+\infty} c_n x^n,$$

Table 4.12: Comparison of the exact solution (4.14) with the 20th-order approximations of $V(t)$ given by (4.112) when $\hbar = -1/2$ and $\alpha = \pm 1/2, \pm \pi/4$.

t	when $\alpha = -\pi/4$	when $\alpha = -1/2$	when $\alpha=1/2$.	when $\alpha = \pi/4$.	exact-result
1/4	0.2449	0.2449	0.2449	0.2449	0.2449
1/2	0.4621	0.4621	0.4621	0.4621	0.4621
3/4	0.6351	0.6351	0.6351	0.6351	0.6351
1	0.7616	0.7616	0.7616	0.7616	0.7616
3/2	0.9051	0.9051	0.9051	0.9051	0.9051
2	0.9640	0.9640	0.9640	0.9640	0.9640
5/2	0.9866	0.9866	0.9866	0.9866	0.9866
3	0.9951	0.9951	0.9951	0.9951	0.9951
4	0.9993	0.9993	0.9993	0.9993	0.9993
5	0.9999	0.9999	0.9999	0.9999	0.9999
10	1.0000	1.0000	0.9999	1.0000	1.0000
100	1.0000	1.0000	1.0000	1.0000	1.0000

corresponds to $[m, n]$ The approximant of Padé is represented as

$$\frac{\sum_{k=0}^{m} a_{m,k}\, x^k}{\sum_{k=0}^{n} b_{m,k}\, x^k},$$

where the coefficients $c_j (j = 0, 1, 2, 3, \ldots, m + n)$ determine $a_{m,k}, b_{m,k}$. The convergence region and rate of a given series can frequently be significantly increased by using the traditional Padé technique. For instance, by using the traditional Padé approach to the perturbation series (4.16), we have the Padé approximants $[1, 1], [2, 2]$ and $[3, 3]$.

$$t, \quad \frac{3t}{3 = t^2}, \quad \frac{t(15 + t^2)}{15 + 6t^2},$$

respectively. The $[m, m]$ Padé approximant can be stated generally as

$$\frac{\sum_{n=0}^{m} a_0^{m,n}\, t^n}{\sum_{n=0}^{m-1} b_0^{m,n}\, t^n}, \qquad \text{when } m \text{ is an odd number,} \qquad (4.113)$$

or

$$\frac{\sum_{n=0}^{m-1} a_0^{m,n}\, t^n}{\sum_{n=0}^{m} b_0^{m,n}\, t^n}, \qquad \text{when } m \text{ is an even number,} \qquad (4.114)$$

where $a_0^{m,n}$ and $b_0^{m,n}$ are coefficients. Take note that as t increases, all of these traditional Padé approximants tend to either infinity or zero. The perturbation solution (4.16)'s $[4,4]$ and $[10,10]$ Padé approximants are depicted in Figure 4.9.

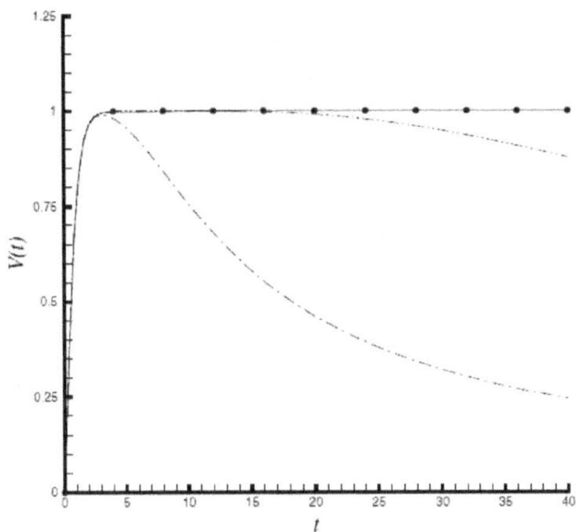

Figure 4.9: Comparison of the exact solution (4.14) with the homotopy-Padé approximation (4.117) and traditional Padé approximant (4.114) of $V(t)$. Symbols: exact solution; solid line: $[4,4]$ homotopy-Padé approximant of $V(t)$; dashdotted line: $[4,4]$ traditional Padé approximant of $V(t)$; dash-dot-dotted line: $[10,10]$ traditional Padé approximant.

By integrating the traditional Padé approach discussed above with the homotopy analysis method, the so-called homotopy-Padé technique was proposed. We first use the traditional $[m,n]$ Padé technique about the embedding parameter q to generate the $[m,n]$ Padé approximant to make sure that the series (4.43) is convergent at $q = 1$.

$$\frac{\sum_{k=0}^{m} A_{m,k}(t)\, q^k}{\sum_{k=0}^{n} B_{m,k}(t)\, q^k}, \tag{4.115}$$

where the first few approximations determine the coefficients $A_{m,k}(t)$ and $B_{m,k}(t)$

$$V_0(t), V_1(t), V_2(t), \ldots, V_{m+n}(t).$$

Then, using (4.39) and $q = 1$ in (4.115), we have what is known as the $[m,n]$ homotopy-Padé approximant.

$$\frac{\sum_{k=0}^{m} A_{m,k}(t)}{\sum_{k=0}^{n} B_{m,k}(t)}, \tag{4.116}$$

The coefficients $A_{m,n}(t)$ and $B_{m,n}(t)$ for the example shown depend on the base functions used to provide the solution $V(t)$. We have the matching $[1,1]$ using the base functions denoted by (4.73). homotopy-Padé approximant.

$$\frac{t(12 + 16t + 7t^2)}{(1+t)(12 + 4t + 7t^2)}$$

and the $[2,2]$ homotopy-Padé approximant

$$\frac{t(168000 + 362880\,t + 238000\,t^2 + 14160\,t^3 - 47124\,t^4 - 36308\,t^5 - 13419\,t^6)}{3(1+t)(56000 + 64960\,t + 33040\,t^2 + 12000\,t^3 - 2508\,t^4 - 9076\,t^5 - 4473\,t^6)}$$

respectively. The $[m,m]]$ homotopy-Padé approximation is typically represented as

$$\frac{\sum_{n=1}^{m^2+m+1} a_2^{m,n}\, t^n}{\sum_{n=0}^{m^2+m+1} b_2^{m,n}\, t^n}, \tag{4.117}$$

where the coefficients are $a_2^{m,n}$ and $b_2^{m,n}$. The discovery that $a_2^{m,n}$ and $b_2^{m,n}$ are independent of the auxiliary parameter $\hbar$ is quite intriguing. We discover that the $[m,m]$ homotopy-Padé approximant is similar to the conventional $[m^2+m+1,\ m^2+m+1]$ Padé approximant in terms of accuracy when comparing (4.117) with (4.113) and (4.114). In contrast to the traditional Padé approximants (4.113) and (4.114), where all correctly go to 1 as $t \to +\infty$, unlike the homotopy-Padé approximant (4.117), which tend to either infinity or zero as $t \to +\infty$. So, compared to the conventional $[m,m]$ Padé approximants ((4.113) and (4.114)), the $[m,m]$ homotopy-Padé approximant (4.117) is substantially more accurate for a given m. For instance, as shown in Figure 4.9, the $[4,4]$ homotopy-Padé approximant is more precise and superior to the $[4,4]$ traditional Padé approximant and even outperforms the $[10,10]$ traditional Padé approximant. In particular, we have the $[1,1]$ homotopy-Padé approximant using the base functions given by (4.81).

$$\frac{1 - \exp(-2t)}{1 + \exp(-2t)} \tag{4.118}$$

It is the exact solution $V(t) = \tanh(t)$. As a result, the so-called homotopy Padé method is significantly more effective than the traditional Padé method.

Similar to this, the related series' convergence can be speed up using the so-called homotopy-Padé technique. For instance, in order to speed up the series of $V''(0)$ and $V'''(0)$, we first use the traditional Padé approach on the series

$$\frac{\partial^2 \Phi(t; q)}{\partial t^2}\Big|_{t=0} = \sum_{n=0}^{+\infty} V_n''(0)q^n$$

and

$$\frac{\partial^3 \Phi(t; q)}{\partial t^3}\Big|_{t=0} = \sum_{n=0}^{+\infty} V_n'''(0)q^n$$

Set $q = 1$ to obtain the corresponding $[m, n]$ homotopy Padé approximants after obtaining their respective $[m, n]$ Padé approximants around the embedding parameter q. Table [4.13] lists the homotopy-Padé approximations of $V''(0)$ and $V'''(0)$, which correspond to the solution expression (4.80) given by the fractional functions. Table [4.14] lists the homotopy-Padé approximations of $V''(0)$ and $V'''(0)$ that corresponds to the solution expression (4.4.5) given by the exponential functions. The homotopy-Padé technique significantly speeds up the convergence of $V''(0)$ and $V'''(0)$ in both situations.

Table 4.13: The $[m, m]$ homotopy-Padé approximation of $V''(0)$ and $V'''(0)$ corresponding to (4.80).

[m,m]	$V''(0)$	$V'''(0)$
$[1, 1]$	0	-3
$[2, 2]$	0	-2
$[3, 3]$	0	-2
$[4, 4]$	0	-2
$[5, 5]$	0	-2
$[10, 10]$	0	-2

For the example case, it is discovered that all of the $[m, m]$ homotopy-Padé approximants are independent of the auxiliary parameter $\hbar$. Therefore, we can still use the homotopy-Padé technique to obtain a convergent result even if we choose a poor value of $\hbar$ such that the related solution series diverges. The $[m, m]$ homotopy-Padé approximants are frequently independent of the auxiliary parameter $\hbar$, as is demonstrated later in this book for various nonlinear situations. But as of right now, we are unable to provide a broad mathematical proof for it.

These examples show how the convergence region and pace of the

Table 4.14: The $[m, m]$ homotopy-Padé approximation of $V''(0)$ and $V'''(0)$ corresponding to (4.4.5).

[m,m]	$V''(0)$	$V'''(0)$
$[1, 1]$	0	-5.57143
$[2, 2]$	0	-2
$[3, 3]$	0	-2
$[4, 4]$	0	-2
$[5, 5]$	0	-2
$[10, 10]$	0	-2

solution series provided by the homotopy analysis method can be significantly increased using the so-called homotopy-Padé technique.

In conclusion, we discuss the fundamental concepts of the homotopy analysis method using a simple example in this chapter. We demonstrate that the homotopy analysis method consistently produces a variety of solution expressions in the auxiliary parameter $\hbar$, which may be stated by many base functions, in contrast to all previous analytical techniques. We can describe the answer in a variety of base functions using the freedom to choose the initial guess, the auxiliary linear operator, and the auxiliary function. By selecting a better set of base functions, we can approximate nonlinear problems more effectively. To guide the selection of the initial guess, the auxiliary linear operator, and the auxiliary function, the rules of solution expression, coefficient ergodicity, and solution existence are presented. The implementation of the homotopy analysis approach is significantly made simpler by these rules. We show that the auxiliary parameter $\hbar$ can be used to control and alter the convergence region and rate of the solution series. It is simple to determine an appropriate value of $\hbar$ to guarantee that the solution series converge by plotting the so-called $\hbar$-curves. Additionally, it is suggested to use the so-called homotopy-Padé technique, which is frequently considerably more effective than the traditional Padé technique, to quicken the convergence of solution series.

Chapter 5

Homotopy Perturbation Method

Homotopy Perturbation Method (HPM) is combination of homotopy in Topology and classic perturbation technique. It is powerful method to obtain analytic or approximate solution for a wide variety of linear and non linear ordinary and partial differential equations with initial value or initial boundary value problems. Ever since He (1999) used HPM to solve differential equations the technique has been extensively researched and it has further expanded e.g. Homotopy Perturbation Transform Method.

Homotopy Analysis method by S. J. Liao (1997) had provided strong backup to the development of Homotopy Perturbation Method (HPM) by J. H. He (1999).

5.1 Basic Idea of Homotopy Perturbation Method (HPM)

Consider the following general form of the time-dependent differential equation to illustrate the fundamental concepts:

$$A(y(r,t)) - f(r,t) = 0, \tag{5.1}$$

where A is a differential operator, $y(r,t)$ is an unknown function, r and t stand for independent variables of space and time, respectively, and $f(r,t)$ is a recognised analytic function. In general, A can be divided into two parts, L and N, where L is a straightforward part that is simple to handle and N contains the remaining elements of A.

$$A = L + N \tag{5.2}$$

Construction of a homotopy $\phi(r, t; q)$ satisfying the following can be done using homotopy technique

$$\begin{aligned} H(\phi(r, t; q), q) &= (1 - q)\{L(\phi(r, t; q)) - L(v_0(r, t))\} \\ &\quad + q\{A(\phi(r, t; q)) - f(r, t)\} \\ &= 0 \end{aligned} \tag{5.3}$$

where $v_0(r, t)$ is an initial guess for Eq. (5.1) that satisfies the initial/boundary condition(s) and $q \in [0, 1]$ is an embedding parameter. Equation (5.3) is known as the homotopy equation. It can also be written as follows:

$$L(\phi(r, t; q)) - L(v_0(r, t)) + q\{N(\phi(r, t; q)) + L(v_0(r, t)) - f(r, t)\} = 0. \tag{5.4}$$

Clearly we have

$$q = 0 \Rightarrow H(\phi(r, t; 0), 0) = L(\phi(r, t; 0)) - L(v_0(r, t)) = 0, \tag{5.5}$$
$$q = 1 \Rightarrow H(\phi(r, t; 1), 1) = A(\phi(r, t; 1)) - f(r, t) = 0, \tag{5.6}$$

This actually has the solution $y(r; t)$ to Eq. (5.1) as the latter. When L is considered to be linear, $v_0(r; t)$ is the only solution to Eq. (5.5), which has $v_0(r; t)$ as one of its solutions. Thus, we have

$$\phi(r, t; 0) = v_0(r, t).$$

$$\phi(r, t; 0) = y(r, t).$$

Deformation is the process of converting q from zero to unity by simply changing $\phi(r; t; q)$ from $v_0(r; t)$ to $y(r; t)$. We can naturally assume that the solution to Equations (5.5) and (5.6) can be expressed as a power series in q, that is the following, if the embedding parameter $q(0 \leq q \leq 1)$ is deemed to be a *small parameter* when using the traditional perturbation technique.i.e.

$$\phi(r, t; q) = u_0(r, t) + u_1(r, t)q + u_2(r, t)q^2 + \ldots \tag{5.7}$$

Using (5.7) for $q = 1$ one has

$$y(r, t) = u_0(r, t) + u_1(r, t) + u_2(r, t) + \ldots \tag{5.8}$$

It is an approximate solution to Eq. (5.1). The series (5.8) is convergent, leading to the exact solution of Eq. (5.1). For approximative answers, one can truncate the series or choose the closed form. The Banach's fixed point theorem can be used to investigate the convergence of the series (5.8) because this method is iterative.

5.2 Guidelines for choosing homotopy equation

We are primarily concerned with the auxiliary operator L and the initial guess v_0 in a homotopy equation. The homotopy equation is totally predetermined once one selects these components because the remaining portion is essentially the original equation, which is less flexible, and is hence less subject to modification. Here, we go over a few broad guidelines for selecting L and v_0.

The homotopy perturbation approach states that L should be:

(i) "Easy to handle".
We mean that it must be selected in a way that makes it easy to go on and solve systems of the resulting equations. It should be emphasized that this requirement does not impose a linear constraint on L. A nonlinear selection of L might be more appropriate in some situations, as He showed while solving the Lighthill equation. But taking a linear operator as L is strongly advised for beginners.

(ii) Closely related to the original equation".
In order to construct L correctly, it is preferable to use a part of the original equation. The effectiveness of this viewpoint is demonstrated by Chowdhury and Hashimt that achieved excellent results by properly selecting the L component (in a later section, we will demonstrate that they could have selected different operators as L).

Although there is no one way for selecting the initial guess in iterative approaches, we can draw the following conclusions from prior research on HPM and our personal experiences:

(i) "It should be obtained from the original equation".
For instance, it might be selected from the initial or boundary conditions or as the answer to a particular component of the original problem.

(ii) "It should reduce complexity of the resulting equations".
Many authors have utilized this criterion when they came across various options as an initial guess, even though it can only be verified after solving some of the resulting system's initial few equations.

5.3   The classic view on HPM

As suggested by several writers, in order to get a good approximation, one must test various options for L and v_0 before selecting the best one. There is no general rule to choosing L and v_0, however, this is the traditional method of employing HPM. In order to review this classic viewpoint, we now give two comparable cases.

Example 1: Consider the time-dependent Emden-Fowler equation

$$y_{xx} + \frac{2}{x}y_x - (6 + 4x^2 - \cos(t))y = y_t,$$

With $y(x,0) = e^{x^2}$ as the initial condition, and $y(0,t) = e^{\sin(t)}$ and $y_x(0,t) = 0$, as the boundary conditions.

(i) We choose $L\phi = \dfrac{\partial^2 \phi}{\partial x^2} + \dfrac{2}{x}\dfrac{\partial \phi}{\partial x}$ and $v_0(x,t) = y(0,t) = e^{\sin(t)}$ this gives the homotopy equation as follows:

$$\phi_{xx} + \frac{2}{x}\phi_x - v_{0x} + q\left\{v_{0xx} + \frac{2}{x}v_{0x} - (6 + 4x^2 - \cos(t))\phi - \phi_t\right\} = 0$$

The following equation system results from applying equation 5.7 and then equating the terms using identical powers of q.

$$u_{0xx} + \frac{2}{x}u_{0x} - v_{0xx} - \frac{2}{x}v_{0x} = 0,$$

$$u_0(0,t) = e^{\sin(t)}, u_{0x}(0,t) = 0,$$

$$u_{1xx} + \frac{2}{x}u_{1x} + v_{0xx} + \frac{2}{x}v_{0x} - (6 + 4x^2 - \cos(t))u_0 - u_{0t} = 0,$$

$$u_1(0,t) = 0, u_{1x}(0,t) = 0,$$

$$u_{2xx} + \frac{2}{x}u_{2x} - (6 + 4x^2 - \cos(t))u_1 - u_{1t} = 0,$$

$$u_2(0,t) = 0, u_2x(0,t) = 0,$$

$$\vdots$$

Subsequently solving the above equations we have

$$u_0(x,t) = e^{\sin(t)},$$

$$u_1(x,t) = e^{\sin(t)}\left\{x^2 + \frac{1}{5}x^4\right\},$$

$$u_2(x,t) = e^{\sin(t)}\left\{\frac{3}{10}x^4 + \frac{13}{105}x^6 + \frac{1}{90}x^8\right\},$$

$$u_3(x,t) = e^{\sin(t)}\left\{\frac{3}{70}x^6 + \frac{17}{360}x^8 + \frac{59}{11550}x^{10} + \frac{1}{3510}x^{12}\right\},$$

$$\vdots$$

Finally, the approximate solution in a series form, according to (5.8) is

$$y(x,t) = e^{\sin(t)}\left\{1 + x^2 + \frac{x^4}{2!} + \frac{x^6}{3!} + \frac{x^8}{4!} + \cdots\right\},$$

leading to the closed form $y(x;t) = e^{\sin(t)+x^2}$, which is the exact solution.

(ii) We choose $L\phi = \dfrac{2}{x}\dfrac{\partial \phi}{\partial x}$ and $v_0(x,t) = 0$.

So we have the homotopy equation as follows:

$$\frac{2}{x}\phi_x + q[\phi_{xx} - (6 + 4x^2 - \cos(t))\phi - \phi_t] = 0.$$

Tthen equating the terms with identical powers of q, we have the following system of equations

$$\frac{2}{x}u_{0x} = 0, \qquad\qquad u_0(0,t) = e^{\sin(t)},$$

$$\frac{2}{x}u_{1x} - (6 + 4x^2 - \cos(t))u_0 - u_{0t} = 0, \qquad\qquad u_1(0,t) = 0,$$

$$\frac{2}{x}u_{2x} - (6 + 4x^2 - \cos(t))u_1 - u_{1t} = 0, \qquad\qquad u_2(0,t) = 0,$$

$$\vdots \qquad\qquad\qquad\qquad \vdots$$

which yields to

$$u_0(x,t) = e^{\sin(t)},$$

$$u_1(x,t) = e^{\sin(t)}\left\{\frac{3}{2}x^2 + \frac{1}{2}x^4\right\},$$

$$u_2(x,t) = e^{\sin(t)}\left\{\frac{-3}{4}x^2 + \frac{3}{8}x^4 + \frac{9}{12}x^6 + \frac{1}{8}x^8\right\},$$

$$u_3(x,t) = e^{\sin(t)}\left\{\frac{3}{8}x^2 - \frac{9}{8}x^4 - \frac{93}{48}x^6 - \frac{1}{16}x^8 + \frac{15}{80}x^10 + \frac{1}{48}x^12\right\},$$

$$\vdots$$

We have

$$y(x,t) = e^{\sin(t)}\left\{1 + 3\left(\frac{1}{2} - \frac{1}{4} + \frac{1}{8} - \frac{1}{16} + \dots\right)x^2 + \left(\frac{1}{2} + \frac{3}{8} - \frac{9}{8} + \dots\right)x^4 + \dots\right\},$$

which yields $y(x,t) = e^{\sin(t)}\left\{1 + x^2 + \frac{x^4}{2!} + \dots\right\}$, leading to the closed

form $y(x,t) = e^{\sin(t)+x^2}$, which is also the exact solution.

(iii) Our choice $L\phi = \dfrac{\partial\phi}{\partial x}$ and $v_0(x,t) = 0$.

So we have the homotopy equation as follows:

$$\phi_t + q\left\{\phi_{xx} - \frac{2}{x}\phi_x + (6 + 4x^2 - \cos(t))\phi\right\} = 0.$$

Then equating the terms with identical powers of q, and then solving the resulting equations, one has

$$u_0(x,t) = e^{x^2},$$

$$u_1(x,t) = \sin(t)e^{x^2},$$

$$u_2(x,t) = \frac{1}{2}\sin^2(t)e^{x^2},$$

$$u_3(x,t) = \frac{1}{6}\sin^3(t)e^{x^2},$$

$$\vdots$$

Again employing (5.8) the approximate solution is

$$y(x,t) = e^{x^2}\left\{1 + \sin(t) + \frac{\sin^2(t)}{2!} + \frac{\sin^3(t)}{3!} + \dots\right\},$$

leading to the closed form $y(x,t) = e^{\sin(t)+x^2}$, which is also the exact solution.

Example 2: Consider the Cauchy reaction-diffusion equation

$$y_t = y_{xx} - y,$$

where the initial condition is that $y(x,o) = e^{-x} + x$ and the boundary conditions are that $y(0,t) = 1$ and $y_x(0,t) = e^{-t} - 1$
. (i) Our choice $L\phi = \dfrac{\partial \phi}{\partial x}$ and $v_0(x,t) = 0$.
Thus, the following homotopy equation exists.

$$\phi_t + q\{-\phi_{xx} + \phi\} = 0.$$

The system of equations that results by equating the terms with the same power of q, and then solving the resulting equation system

$$u_0(x,t) = e^{-x} + x,$$
$$u_1(x,t) = -xt,$$
$$u_2(x,t) = x\frac{t^2}{2!},$$
$$u_3(x,t) = -x\frac{t^3}{3!},$$
$$\vdots$$

So we have the approximate solution

$$y(x,t) = e^{-x} + x\left(1 - t + \frac{t^2}{2!} - \frac{t^3}{3!} + \cdots\right),$$

yielding the exact answer in the closed form $y(x,t) = e^{-x} + xe^{-t}$.

(ii) We select $L\phi = \dfrac{\partial \phi}{\partial x}$ and $v_0(x,t) = y(x,0) = e^{-x} + x$; take note that in this case, we have placed our initial guess as the equation's initial condition. We end up with the identical phrases as in (i).

(iii) We decide on $L\phi = \dfrac{-\partial^2 \phi}{\partial x^2}$ and $v_0(x,t) = 0$.
Thus, we have the homotopy equation

$$-\phi_{xx} + q\{\phi_t + \phi\} = 0.$$

We have

$$u_0(x,t) = 1 + x(e^{-t}11),$$

$$u_1(x,t) = \frac{t^2}{2!} - \frac{t^3}{3!},$$

$$u_2(x,t) = \frac{t^4}{4!} - \frac{t^5}{5!},$$

$$\vdots$$

Leading to the exact solution

$$y(x,t) = xe^{-t} + \left(1 - x + \frac{x^2}{2!} - \frac{x^3}{3!} + \frac{x^4}{4!} - \frac{x^5}{5!} + \dots\right) = xe^{-t} + e^{-x}.$$

(iv) We decide on $L\phi = \dfrac{-\partial^2\phi}{\partial x^2}$ and

$$v_0(x,t) = y(0,t) + xy_x(0,t) = 1 + x(e^{-t} - 1);$$

in this case, we constructed an initial guess using the problem's boundary conditions. In this instance, we reach the terms acquired in (iii) precisely.

The effectiveness and adaptability of HPM in solving equations are demonstrated by these comparative examples. It is important to keep in mind that HPM has several drawbacks, such as when the studied equation involves terms like $ln(y), \sin(y), e^y, \cosh(y), \dots$ where y is the unknown function. To achieve an analytical or approximate answer, we must employ Taylor's expansion or other approximations in these circumstances before applying HPM.

5.4   Proposed choices for L and v_0

We suggest selecting the L component of time-dependent differential equations using the highest order of derivative with respect to (only) t, i.e.

$$L\phi = \frac{\partial\phi}{\partial t}, \qquad\qquad L\phi = \frac{\partial^2\phi}{\partial t^2}, \dots$$

depending on the appearance of them in the equation under consideration.

Also, we suggest to put $v_0 = 0$; i.e. the zero function.

Derivatives with regard to time that have the simple forms $y_t, y_{tt}, \dots$ (not depending on the spatial variables of the equation) are always present in time-dependent differential equations. Consequently, in light of the explanation in Section 3.1, the L part can be selected as

$$L\phi = \frac{\partial \phi}{\partial t}$$

or

$$L\phi = \frac{\partial^2 \phi}{\partial t^2}, \dots.$$

Our first equation, for example, would be $u_{0t} = 0$ with the initial condition being $u_0(x,0) = y(x,0)$, when we establish the first guess (v_0) as the zero function by applying equation (5.7).

The solution to this equation is always $u_0(x,t) = y(x,0)$. We immediately allow the beginning conditions of the issue to play their part and effectively participate in the development of the solution because $u_0(x,t)$ is the first term of our approximation of equation 5.8. Consider the Cauchy reaction-diffusion equation in Example 4.2(i) as an illustration. If we take the initial guess to be $v_0(x,t) = y(0,t) + y_x(0,t)$, we can't readily solve the ensuing equations, and the HPM fails to solve the equation. Subsequently we apply the rules suggested in Section 3.2 by selecting $v_0 = 0$. Additionally, by doing it this manner, we avoid adding additional terms that might possibly undermine our analytic approximation (this comment is quite heuristic).

5.5   Examples:

5.5.1   Evolution equations

Utilising $L\phi = \dfrac{\partial \phi}{\partial t}$, Ganji et al. have solved three examples of evolution equations via HPM. and using the basic hypothesis as the initial point. These three cases are presented here, and they are resolved by selecting $L\phi = \dfrac{\partial \phi}{\partial t}$ and $v_0(x,t) = 0$ as the initial hypothesis.

Example 1:

Take into consideration the initial condition $y_t - y_{xx} + \left(\dfrac{y^2}{2}\right)_x = 0$ with the initial condition $y(x,0) = x$. The homotopy equation is

$$\phi_t + q\left\{ -\phi_{xxt} + \left(\frac{\phi^2}{2}\right)_x \right\} = 0.$$

Using equation (5.7), equating the terms using their corresponding powers of q, and then solving the resulting equations, one has

$$u_0(x,t) = x,$$
$$u_1(x,t) = -xt,$$
$$u_2(x,t) = xt^2,$$
$$u_3(x,t) = -xt^3,$$

$$\vdots$$

Thus, we have the following approximation to the solution:

$$y(x,t) = x(1 - t + t^2 - t^3 + \ldots) = \frac{x}{1+t}.$$

which is less complicated.

Example 2:

Consider the equation $y_t + y_x = 2y_{xxt}$ with the initial condition $y(x,o) = e^{-x}$. we have following homotopy

$$\phi_t + q\{\phi_x - 2\phi_{xxt}\} = 0.$$

Using (5.7), equating the terms using their corresponding powers of q, and then solving the resulting equations, one has

$$u_0(x,t) = te^{-x},$$
$$u_1(x,t) = \left(\frac{t^2}{2} + 2t\right)e^{-x},$$
$$u_2(x,t) = \left(\frac{t^3}{3!} + 2t^2 + 4t\right)e^{-x},$$
$$u_3(x,t) = \left(\frac{t^4}{4!} + t^3 + 6t^2 + 8t\right)e^{-x},$$

$$\vdots$$

so the 7-term approximation is

$$\left(1 + 63t + \frac{129}{2}t^2 + \frac{37}{2}t^3 + \frac{49}{24}t^4 + \frac{11}{120}t^5 + \frac{1}{720}t^6\right)e^{-x}$$

It is the same approximation found using the HPM and the variational iteration method (VIM).

Example 3:

Take into account the initial condition $y(x; 0) = \sin(x)$ for the equation $y_t - y_{xxxx} = 0$. The following homotopy equation is

$$\phi_t + q\{-\phi_{xxxx}\} = 0.$$

Using (5.7), then equating the terms with identical powers of q, then solving the resulting system of equations one can see

$$u_0(x, t) = \sin(x),$$
$$u_1(x, t) = -t\sin(x),$$
$$u_2(x, t) = \frac{t^2}{2!}\sin(x),$$
$$u_3(x, t) = \frac{t^3}{3!}\sin(x),$$
$$\vdots$$

Thus, we have the following approximation to the solution:

$$y(x, t) = \sin(x)\left(1 - t + \frac{t^2}{2!} - \frac{t^3}{3!} + \dots\right) = \frac{\sin(x)}{1 + t},$$

which is the exact solution and whose closed form is significantly simpler.

5.5.2   Cauchy reaction-diffusion equations

Different approaches have been used to solve Cauchy reaction-diffusion equations . We solved one such equation in Example 4.2. Here, we give a few more examples from this family to compare our approximations with the results from HAM.

Example 1

We take into account the equation $y_t = y_{xx} + 2ty$ with the condition $y(x, 0) = e^x$. The following homotopy equation is

$$\phi_t + q - \phi_{xx} - 2\phi t = 0.$$

Applying (5.7), equating terms with the same power of q, and solving the resulting system of equations gives us the following,

$$u_0(x,t) = e^x,$$

$$u_1(x,t) = (t+t^2)e^x$$

$$u_2(x,t) = \frac{(t+t^2)^2}{2!}e^x,$$

$$u_3(x,t) = \frac{(t+t^2)^3}{3!}e^x,$$

$$\vdots$$

Consequently, we have the following as an approximate solution:

$$y(x,t) = e^{x^2}\left(1 + t + t^2 + \frac{(t+t^2)^2}{2!} + \frac{(t+t^2)^3}{3!} + \cdots\right),$$

giving the exact solution derived using HAM in the closed form

$$y(x,t) = e^{x+t+t^2}$$

.

Example 2

The initial condition is $y(x,0) = e^{x^2}$ and the equation is $y_t = y_{xx} - (1 + 4x^2)y$. The homotopy equation is as follows,

$$\Phi_t + q\{-\phi_{xx} + (1+4x^2)\phi\} = 0$$

One can see the solution by using (5.7), equating the terms with identical powers of q, and then solving the resulting system of equations.

$$u_0(x,t) = e^{x^2},$$

$$u_1(x,t) = e^{x^2}t,$$

$$u_2(x,t) = e^{x^2}\frac{t^2}{2!},$$

$$u_3(x,t) = e^{x^2}\frac{t^3}{3!},$$

$$\vdots$$

Consequently, we have the following as an approximate solution:

$$y(x,t) = e^{x^2}\left(1 + t + \frac{t^2}{2} + \frac{t^3}{3!} + \cdots\right),$$

yielding the closed form $y(x,t) = e^{x^2+t}$, which is the exact solution.

Example 3

With the initial condition $y(x,0) = e^{x^2}$, take a look at the equation $y_t = y_{xx} - (4x^2 - 2t + 2)y$. The homotopy equation is as follows,

$$\phi_t + q - \phi_{xx} + (4x^2 - 2t + 2)\phi = 0$$

. Using (5.7) and solving the resulting system of equations after equating the terms with the same powers of q, we have

$$u_0(x,t) = e^{x^2},$$
$$u_1(x,t) = e^{x^2} t^2,$$
$$u_2(x,t) = e^{x^2} \frac{t^4}{2!},$$
$$u_3(x,t) = e^{x^2} \frac{t^6}{3!},$$
$$\vdots$$

So we have the approximate solution as follows:

$$y(x,t) = e^{x^2} \left(1 + t^2 + \frac{t^4}{2!} + \frac{t^6}{3!} + \cdots \right),$$

yielding the closed form $y(x,t) = e^{x^2 + t^2}$, which is the exact solution.

5.5.3 Emden-Fowler type equations

Example 1

Take into account the initial condition $y(x,0) = x^2 + e^{x^2}$ and the equation $y_{xx} + \frac{2}{x} y_x - (5 + 4x^2)y = y_t + (6 - 5x^2 - 4x^4)$. We have the following homotopy equation after using our method.

$$\phi_t + q \left\{ -\phi_{xx} - \frac{2}{x}\phi_x + (5 + 4x^2)\phi + (6 - 5x^2 - 4x^4) \right\} = 0.$$

Applying formula (5.7), equating terms whose powers of q are the same, and then solving the resulting equations gives us

$$u_0(x,t) = x^2 + e^{x^2},$$

$$u_1(x,t) = e^{x^2}t,$$

$$u_2(x,t) = e^{x^2}\frac{t^2}{2!},$$

$$u_3(x,t) = e^{x^2}\frac{t^3}{3!},$$

$$\vdots$$

So according to (5.8), the approximate solution is

$$y(x,t) = x^2 + e^{x^2}\left(1 + t + \frac{t^2}{2!} + \frac{t^3}{3!} + \cdots\right),$$

resulting in the exact solution, which has less complexity, in the closed form $y(x,t) = x^2 + e^{t+x^2}$.

Example 2

Take the equation $y_{xx} + \dfrac{2}{x}y_x - (5 + 4x^2)y = y_{tt} + (12x - 5x^3 - 4x^5)$ with the initial condition $y(x,0) = x^3 + e^{x^2}$ into consideration. We have the following homotopy equation after using our method.

$$\phi_{tt} + q\left\{-\phi_{xx} - \frac{2}{x}\phi_x + (5 + 4x^2)\phi + (12 - 5x^3 - 4x^5)\right\} = 0.$$

Using (5.7) and solving the resulting system of equations after equating the terms with the same powers of q, we have

$$u_0(x,t) = x^3 + (1 - t)e^{x^2},$$

$$u_1(x,t) = e^{x^2}\left(\frac{t^2}{2!} - \frac{t^3}{3!}\right),$$

$$u_2(x,t) = e^{x^2}\left(\frac{t^4}{4!} - \frac{t^5}{5!}\right),$$

$$\vdots$$

So according to (5.8), the approximate solution is

$$y(x,t) = x^3 + e^{x^2}\left(1 - t + \frac{t^2}{2!} - \frac{t^3}{3!} + \frac{t^4}{4!} - \frac{t^5}{5!} + \cdots\right),$$

leading to the closed form $y(x,t) = x^3 + e^{x^2 - t}$, which is the exact solution.

Example 3

With the initial conditions $y(x,0) = e^{x^3}$ and $y_t(x,0) = 0$, take a look at the equation $y_{xx} + \dfrac{4}{x} y_x - (18 + 9x^4)y = y_{tt} - 2 - (18x + 9x^4)t^2$

In this example we use the modified HPM, proposed by Odibat, so we have the modified homotopy equation as follows:

$$\phi_{tt} + q\left\{ -\phi_{xx} - \frac{4}{x}\phi_x + (18x + 9x^4)\phi + (18x + 9x^4)t^2 \right\} = 2.$$

Using (5.7) and solving the resulting system of equations after equating the terms with the same powers of q, we have

$$u_0(x,t) = t^2 + e^{x^3},$$
$$u_1(x,t) = 0,$$
$$u_2(x,t) = 0,$$
$$u_3(x,t) = 0,$$
$$\vdots$$

Employing equation (5.8), we simply have the exact solution

$$y(x,t) = t^2 + e^{x^3},$$

5.5.4 Klein-Gordon equations

Example 1

We take into account the linear Klein-Gordon equation $y_{tt} - y_{xx} = y$ with the initial conditions $y(x,0) = 1 + \sin(x)$

The following homotopy equation results from our selection,

$$\phi_{tt} + q\{-\phi_{xx} - \phi\} = 0.$$

Using equation (5.7) and solving the resulting system of equations after equating the terms with the same powers of q, we have

$$u_0(x, t) = 1 + \sin(x),$$

$$u_1(x, t) = \frac{t^2}{2!},$$

$$u_2(x, t) = \frac{t^4}{4!},$$

$$u_3(x, t) = \frac{t^6}{6!},$$

$$\vdots$$

So according to equation (5.8), the approximate solution is

$$y(x, t) = \sin(x) + \left(1 + \frac{t^2}{2!} + \frac{t^4}{4!} + \frac{t^6}{6!} + \cdots\right),$$

which results in $y(x, t) = \sin(x) + \cosh(t)$, the closed form.

Example 2

We look at the nonlinear nonhomogeneous Klein-Gordon equation

$$y_{tt} - y_{xx} - 2y = -2\sin(x)\sin(t)$$

with the initial conditions $y(x, 0) = 0$ and $y_t(x, 0) = \sin(x)$. Our decision results in the homotopy equation

$$\phi_{tt} + q\left\{-\phi_{xx} - 2\phi + 2\sin(x)\sin(t)\right\} = 0.$$

Using equation (5.7) and solving the resulting system of equations after equating the terms with the same powers of q, we have

$$u_0(x, t) = t\sin(x)$$

$$u_1(x, t) = \sin(x)\left\{2\sin(t) - 2t + \frac{t^3}{6}\right\}$$

$$u_2(x, t) = \sin(x)\left\{-2\sin(t) + 2t - \frac{t^3}{3} + \frac{t^5}{120}\right\}$$

$$u_3(x, t) = \sin(x)\left\{+2\sin(t) - 2t + \frac{t^3}{3} - \frac{t^5}{60} + \frac{t^7}{7!}\right\}$$

$$u_4(x, t) = \sin(x)\left\{-2\sin(t) + 2t - \frac{t^3}{3} + \frac{t^5}{60} - \frac{t^7}{2 \times 6!} + \frac{t^9}{9!}\right\}$$

In order to acquire an approximate solution , a few terms from u_{2i-1} are cancelled out with terms from u_{2i}, and then an approximate solution is,

$$y(x,t) = \sin(x)\left\{ t - \frac{t^3}{3!} + \frac{t^5}{5!} - \cdots \right\}$$

This indicates that $y(x,t) = \sin(x)\sin(t)$ is the closed form. The above-mentioned equation's exact solution was discovered by Chowdhury and Hashim after only one iteration. Due to the fact that we use a general approach, our option also produces a convergent series, which is noteworthy.

Example 3

We take into account the nonlinear nonhomogeneous Klein-Gordon equation

$$y_{tt} - y_{xx} + y^2 = -x\cos(t) + x^2\cos^2(t)$$

with the initial conditions $y(x,0) = x$ and $y_t(x,0) = 0$. This equation has the exact solution $y(x,t) = x\cos(t)$. Despite appearing difficult, our analytical approximation converges to the exact solution. The approximate 4-term is

$$y_{\text{app }4}(x,t) = u_0(x,t) + u_1(x,t) + u_2(x,t) + u_3(x,t)$$

$$= \cos(t)\left\{ \frac{1}{32}t^2\cos^2(t) + \frac{5}{192}t^4 - \frac{33}{256}\cos^2(t) - \frac{23}{256}t^2 + \frac{1}{256}\cos^4(t) + \frac{1}{8} \right.$$

$$\left. - \frac{7}{1440}t^6 - \frac{1}{16}t\cos(t)\sin(t) \right\}x^4$$

$$+ \cos(t)\left\{ \frac{-64}{9} + 2t\sin(t) + \frac{43}{6}\cos(t) + 2t^2 - \frac{1}{6}t^4 - \frac{1}{2}t^2\cos(t) - \frac{1}{18}\cos^3(t) \right\}x^3$$

$$+ \cos(t)\left\{ \frac{1}{4}\cos^2(t) + \frac{1}{4}t^2 - \frac{1}{12}t^4 - \frac{1}{4} + \cos(t) + \frac{1}{90}t^6 \right\}x$$

$$+ \cos(t)\left\{ \frac{1}{8}\cos^2(t) - \frac{15}{8}t^2 - 4\cos(t) + \frac{1}{8}t^4 + \frac{31}{8} \right\}$$

We computed the absolute errors for this approximation in Table 5.1, demonstrating the effectiveness of our choice.

It appears that as t and x are increased, the inaccuracy grows as well. It is as a result of the fact that we approximated using only 4 terms. So, it isn't a major issue. One should include more terms in their estimate in order to acquire findings that are more accurate (with reduced error values).

Table 5.1: Absolute errors of a 4-term approximation of Example 3

x_i	t_i	$|y - y_{app4}|$
0.1	0.1	3.749×10^{-12}
0.2	0.2	$1 : 128 \times 10^{-9}$
0.3	0.3	$3 : 175 \times 10^{-8}$
0.4	0.4	3.229×10^{-7}
0.5	0.5	1.781×10^{-6}

Table 5.2: Absolute errors of a 4-term approximation of Example 4

x_i	t_i	$|y - y_{app4}|$
0.1	0.1	1.247×10^{-16}
0.2	0.2	2.044×10^{-12}
0.3	0.3	5.968×10^{-10}
0.4	0.4	3.349×10^{-8}
0.5	0.5	7.615×10^{-7}

Example 4

Consider the nonlinear nonhomogeneous Klein-Gordon equation

$$y_{tt} - y_{xx} + y^2 = 6xt(x^2 - t^2) + x^6 t^6$$

with the initial conditions $y(x, 0) = 0$ and $y_t(x, 0) = 0$, which has the exact solution $y(x, t) = x^3 t^3$. In this case, our decision results in the following approximate 4-term solution:

$$
\begin{aligned}
y_{app4} &= u_0 + u_1 + u_2 + u_3 \\
&= x^3 t^3 + \frac{53}{4200} x^4 t^{10} - \frac{13}{92400} x^2 t^{12} - \frac{1}{959616} x^{12} t^{18} \\
&\quad + \frac{1}{19600} x^7 t^{15} - \frac{1}{4368} x^9 t^{13}
\end{aligned}
$$

We computed the absolute errors for this approximation in Table 5.2 , demonstrating the effectiveness of our choice.

Once more, by including more terms in our approximation, we can obtain findings that are more accurate for higher values of t and x. Furthermore, we should state that our choice of computing the error in points (x, t) with equal values of x and t is arbitrary, and that other (not equal) values can yield extremely close results.

5.6   Schrodinger Equation

The linear Schrodinger equation is considered as follows:

$$\frac{\partial u}{\partial t} + i\frac{\partial^2 u}{\partial x^2} = 0, \qquad u(x,0) = f(x), \qquad i^2 = -1,$$

according to the He's homotopy perturbation method, we have from above equation

$$(1-p)\left[\frac{\partial u}{\partial t} - \frac{\partial u_0}{\partial t}\right] + p\left[\frac{\partial u}{\partial t} + i\frac{\partial^2 u}{\partial x^2}\right] = 0, \qquad p \in [0,1],$$

or:

$$\frac{\partial u}{\partial t} - \frac{\partial u_0}{\partial t} + p\left[\frac{\partial u_0}{\partial t} + i\frac{\partial^2 u}{\partial x^2}\right] = 0,$$

equation of the form as follows:

$$p^0 : \frac{\partial u_0}{\partial t} - \frac{\partial u_0}{\partial t} = 0,$$

$$p^1 : \frac{\partial u_1}{\partial t} - \frac{\partial u_0}{\partial t} + i\frac{\partial^2 u_0}{\partial x^2} = 0, \qquad u_1(x,0) = 0,$$

$$p^2 : \frac{\partial u_2}{\partial t} + i\frac{\partial^2 u_1}{\partial x^2} = 0, \qquad u_2(x,0) = 0,$$

$$\vdots$$

$$p^k : \frac{\partial u_k}{\partial t} + i\frac{\partial^2 u_{k-1}}{\partial x^2} = 0, \qquad u_k(x,0) = 0$$

$$\vdots$$

The approximate solution Schrodinger equation can be obtained by setting $p = 1$

$$u = \lim U = u_0 + u_1 + u_2 + \dots$$

Example 1:
Consider the following linear Schrodinger equation :

$$\frac{\partial u}{\partial t}(x,t) + i\frac{\partial^2 u}{\partial x^2}(x,t) = 0, \tag{5.9}$$

$$u_0(x) = 1 + \cosh 2x, \qquad x \in \mathbb{R}$$

According to He's homotopy perturbation method, we construct the following homotopy:

$$\frac{\partial u}{\partial t} - \frac{\partial u_0}{\partial t} + p\left[\frac{\partial u_0}{\partial t} + i\frac{\partial^2 u}{\partial x^2}\right] = 0, \qquad (5.10)$$

Let solution be

$$U = u_0 + pu_1 + p^2 u_2 + \dots \qquad (5.11)$$

where p is the embedded parameter $(0 \le p \le 1)$

substituting (5.11) into (5.10), we get following set of differential equations:

$$p^0 : \frac{\partial u_0}{\partial t} - \frac{\partial u_0}{\partial t} = 0, \qquad u_0(x) = 1 + \cosh 2x,$$

$$p^1 : \frac{\partial u_1}{\partial t} - \frac{\partial u_0}{\partial t} + i\frac{\partial^2 u_0}{\partial x^2} = 0, \qquad u_1(x,0) = 0,$$

$$p^2 : \frac{\partial u_2}{\partial t} + i\frac{\partial^2 u_1}{\partial x^2} = 0, \qquad u_2(x,0) = 0,$$

$$\vdots$$

$$(5.12)$$

solving the systems accordingly, thus we obtain:

$$u_0(x) : 1 + \cosh 2x,$$
$$u_1(x) : (-4it)\cosh 2x,$$
$$u_2(x) : \frac{(-4it)^2}{2!}\cosh 2x,$$
$$u_3(x) : \frac{(-4it)^3}{3!}\cosh 2x,$$
$$\vdots$$

By setting $p = 1$, the solution of (5.9) can be obtained as:

$$U = u_0 + u_1 + u_2 + \dots \qquad (5.13)$$

Thus we have:

$$U(x,t) = 1 + \cosh 2x\left(1 + (-4it) + \frac{(-4it)^2}{2!} + \frac{(-4it)^3}{3!} + \dots\right).$$

In the limit of infinitely many terms, yields the closed-form solution:

$$u(x,t) = 1 + e^{-4it}\cosh 2x,$$

which is the exact solution.

Example 2:

Consider the following linear Schrodinger equation :

$$\frac{\partial u}{\partial t}(x,t) + i\frac{\partial^2 u}{\partial x^2}(x,t) = 0, \tag{5.14}$$

$$u_0(x) = e^{3ix}$$

According to He's homotopy perturbation method, we construct the following homotopy:

$$\frac{\partial u}{\partial t} - \frac{\partial u_0}{\partial t} + p\left[\frac{\partial u_0}{\partial t} + i\frac{\partial^2 u}{\partial x^2}\right] = 0, \tag{5.15}$$

substituting (5.11) into (5.15), we get following set of differential equations:

$$\begin{aligned}
p^0 &: \frac{\partial u_0}{\partial t} - \frac{\partial u_0}{\partial t} = 0, & u_0(x) &= e^{3ix} \\
p^1 &: \frac{\partial u_1}{\partial t} - \frac{\partial u_0}{\partial t} + i\frac{\partial^2 u_0}{\partial x^2} = 0, & u_1(x,0) &= 0, \\
p^2 &: \frac{\partial u_2}{\partial t} + i\frac{\partial^2 u_1}{\partial x^2} = 0, & u_2(x,0) &= 0,
\end{aligned} \tag{5.16}$$

$$\vdots$$

solving the systems accordingly, thus we obtain:

$$\begin{aligned}
u_0(x) &: e^{3ix}, \\
u_1(x) &: 9ite^{3ix}, \\
u_2(x) &: \frac{-81}{2}t^2 e^{3ix}, \\
u_3(x) &: \frac{-243}{2}it^3 e^{3ix},
\end{aligned}$$

$$\vdots$$

By setting $p = 1$ in Eq.(5.11), the solution of (5.14) can be obtained as:

$$U = u_0 + u_1 + u_2 + \dots. \tag{5.17}$$

Thus we have:

$$U(x,t) = e^{3ix}\left(1 + (9it) + \frac{(9it)^2}{2!} + \frac{(9it)^3}{3!} + \dots\right).$$

In the limit of infinitely many terms, yields the closed-form solution,

$$u(x,t) = e^{3i(x+3t)},$$

which is the exact solution.

5.7 Nonlinear Partial Differential Equations

Cauchy problem for the nonlinear parabolic–hyperbolic equation of the following type

$$\left(\frac{\partial}{\partial t} - \Delta\right)\left(\frac{\partial^2}{\partial t^2} - \Delta\right) u = F(u),$$

with initial conditions

$$\frac{\partial^k u}{\partial t^k}(0, X) = \varphi_k(X), X = (x_1, x_2, \ldots, x_i), \qquad k = 0, 1, 2.$$

where the nonlinear term is represented by $F(u)$, and Δ is the Laplace operator in $\mathbb{R}_n$.

5.8 Unsteady non-Newtonian Fluid between Stationary and Oscillating Plates

The non-dimensional form of equation and boundary conditions are

$$\frac{\partial v}{\partial t} = \frac{\partial^2 v}{\partial y^2} + \alpha\frac{\partial}{\partial t}\left(\frac{\partial^2 v}{\partial y^2}\right) + 6\beta\left(\frac{\partial v}{\partial y}\right)^2\left(\frac{\partial^2 v}{\partial y^2}\right), \qquad (5.18)$$

Boundary conditions are

$$v(y,t) = \begin{cases} 1 + \cos\omega t & \text{at} \quad y = 1 \\ 0 & \text{at} \quad y = -1 \end{cases} \qquad (5.19)$$

Basics of HPM

To explain the elementary theory of HPM for solving the nonlinear partial differential equation we consider the following general equation

$$\mathcal{R}(v(y,t)) - \mathcal{M}(y,t) = 0, \qquad \mathcal{B}(v(y,t)) = 0, \qquad (5.20)$$

Where $v(y,t)$ is the unknown function, $\mathcal{M}(y,t)$ is the known analytic function, $\mathcal{B}$ is the boundary operator and $\mathcal{R}$ is the general differential operator which is expressed in linear part $\mathcal{L}(v(y,t))$ and nonlinear part $\mathcal{N}(v(y,t))$ as

$$\mathcal{R}(v(y,t)) = \mathcal{L}(v(y,t)) + \mathcal{N}(v(y,t)) \tag{5.21}$$

Therefore equation (5.20) can be written as

$$\mathcal{L}(v(y,t)) + \mathcal{N}(v(y,t)) - \mathcal{M}(y,t) = 0, \tag{5.22}$$

Now according to the homotopic method define as

$$\mathcal{H}(v(y,t),\mathfrak{p}) = (1-\mathfrak{p})[\mathcal{L}(v(y,t)) - \mathcal{L}(v_0(y,t))] \\ + \mathfrak{p}[\mathcal{R}(v(y,t) - \mathcal{M}(y,t))], \tag{5.23}$$

or we can write equation (5.23) as

$$\mathcal{H}(v(y,t),\mathfrak{p}) = \mathcal{L}(v(y,t)) - \mathcal{L}(v_0(y,t)) + \mathfrak{p}\mathcal{L}(v_0(y,t)) \\ + \mathfrak{p}[\mathcal{N}(v(y,t) - \mathcal{M}(y,t))], \tag{5.24}$$

Here $\mathfrak{p} \in [0,1]$ is the embedding parameter and $v_0(y,t)$ is the initial approximation of equation (5.20) satisfying the boundary condition. Now from equation (5.23) and (5.24) we have

$$\mathcal{H}(v(y,t),0) = \mathcal{L}(v(y,t)) - \mathcal{L}(v_0(y,t)) = 0, \tag{5.25}$$

$$\mathcal{H}(v(y,t),1) = \mathcal{R}(v(y,t)) - \mathcal{M}(y,t) = 0, \tag{5.26}$$

By the variation of $\mathfrak{p}$ from 0 to 1, $v(y,t,\mathfrak{p})$ change from $v_0(y,t)$ to $v(y,t)$ which is called Deformation. $\mathcal{L}(v(y,t)) - \mathcal{L}(v_0(y,t))$ and $\mathcal{R}(v(y,t)) - \mathcal{M}(y,t)$ are called Homotropic.

The approximate solution of equation (5.19) can be expressed as a series of the power of p as

$$v(y,t) = v_0(y,t) + \mathfrak{p}v_1(y,t) + \mathfrak{p}^2 v_2(y,t) + \ldots, \tag{5.27}$$

Setting $\mathfrak{p} = 1$, then the approximate solution of equation (5.27) becomes

$$v(y,t) = \lim_{\mathfrak{p}\to 1} v(y,t) = v_0(y,t) + v_1(y,t) + v_2(y,t) + \ldots, \tag{5.28}$$

The HPM Solution of first problem

In this unit we apply the HPM to solve the nonlinear PDE. We obtained the zero, first and second order velocity problems

$$p^0 : \frac{\partial^2 v_0(y,t)}{\partial y^2} = 0 \tag{5.29}$$

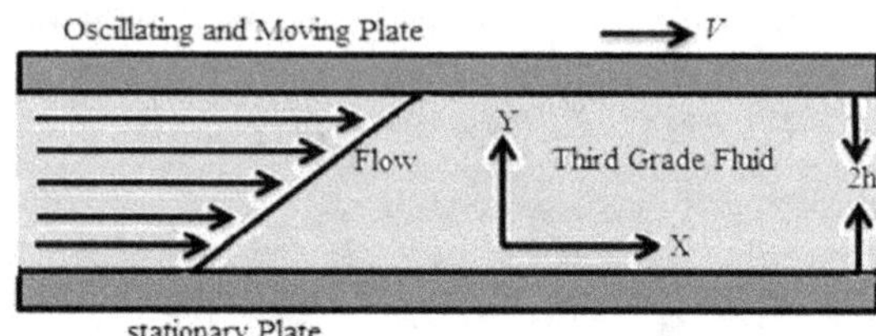

Figure 5.1: Geometry of first problem

$$p^1 : \frac{\partial^2 v_1(y,t)}{\partial y^2} = -\left(\frac{\partial v_0}{\partial t}\right) + 2\frac{\partial^2 v_0}{\partial y^2} + 6\beta\left(\frac{\partial v_0}{\partial t}\right)^2\left(\frac{\partial^2 v_0}{\partial y^2}\right) + \alpha\frac{\partial}{\partial t}\left(\frac{\partial^2 v_0}{\partial y^2}\right), \tag{5.30}$$

$$p^2 : \frac{\partial^2 v_2(y,t)}{\partial y^2} = -\left(\frac{\partial v_1}{\partial t}\right) + 12\beta\left(\frac{\partial v_0}{\partial y}\right)\left(\frac{\partial v_1}{\partial y}\right)\left(\frac{\partial^2 v_0}{\partial y^2}\right) + \frac{\partial^2 v_1}{\partial y^2} + 6\beta\left(\frac{\partial v_0}{\partial t}\right)^2\left(\frac{\partial^2 v_1}{\partial y^2}\right) + \alpha\frac{\partial}{\partial t}\left(\frac{\partial^2 v_1}{\partial y^2}\right) \tag{5.31}$$

Solutions of zero, first and second order problems using boundary conditions in equations (5.18, 5.19) are

$$v_0(y,t) = \cos[\frac{t\omega}{2}]^2 + \cos[\frac{t\omega}{2}]^2 y, \tag{5.32}$$

$$v_1(y,t) = \frac{1}{6}\omega\sin[t\omega]y + \frac{1}{4}\omega\sin[t\omega]y^2 + \frac{1}{12}\omega\sin[t\omega]y^3, \tag{5.33}$$

$$\begin{aligned}
v_2(y,t) = &\left\{\frac{1}{4}\omega^2\cos[t\omega]\left(\frac{1}{12}-\alpha\right) - \frac{3}{32}\beta\omega\csc[\frac{t\omega}{2}]^4\sin[t\omega]^5\right\} \\
&+ \left\{\frac{1}{12}\omega^2\cos[t\omega]\left(\frac{23}{60}-\alpha\right) - \frac{1}{32}\beta\omega\csc[\frac{t\omega}{2}]^4\sin[t\omega]^5\right\}y \\
&+ \left\{\frac{1}{4}\alpha + \frac{3}{32}\beta\omega\csc[\frac{t\omega}{2}]^4\sin[t\omega]^5\right\}y^2 \\
&- \left\{\frac{1}{12}\omega^2\cos[t\omega]\left(\frac{1}{3}-\alpha\right) - \frac{1}{32}\beta\omega\csc[\frac{t\omega}{2}]^4\sin[t\omega]^5\right\}y^3 \\
&- \frac{1}{48}\omega^2\cos[t\omega]y^4 - \frac{1}{240}\omega^2\cos[t\omega]y^5,
\end{aligned} \tag{5.34}$$

General solution of problem is

$$v(y,t) = v_0(y,t) + v_1(y,t) + v_2(y,t), \tag{5.35}$$

$$
\begin{aligned}
v_2(y,t) = &\left\{ \cos[\tfrac{t\omega}{2}]^2 + \tfrac{1}{12}\omega^2 \cos[t\omega]\left(\tfrac{1}{4} - \alpha\right) - \tfrac{3}{32}\beta\omega \csc[\tfrac{t\omega}{2}]^4 \sin[t\omega]^5 \right\} \\
&+ \left\{ \cos[\tfrac{t\omega}{2}]^2 + \tfrac{1}{12}\omega^2 \cos[t\omega]\left(\tfrac{23}{60} - \alpha\right) + \tfrac{1}{6}\omega \sin[t\omega] \right. \\
&\left. - \tfrac{3}{32}\beta\omega \csc[\tfrac{t\omega}{2}]^4 \sin[t\omega]^5 y \right. \\
&+ \left\{ \tfrac{1}{4}\alpha\omega^2 \cos[t\omega] + \tfrac{1}{4}\omega \sin[t\omega] + \tfrac{3}{32}\beta\omega \csc[\tfrac{t\omega}{2}]^4 \sin[t\omega]^5 \right\} y^2 \\
&+ \left\{ \tfrac{1}{12}\omega^2 \cos[t\omega]\left(\alpha - \tfrac{1}{3}\right) + \tfrac{1}{12}\beta\omega \csc[\tfrac{t\omega}{2}]^4 \sin[t\omega]^5 \right\} y^3 \\
&- \tfrac{1}{48}\omega^2 \cos[t\omega]y^4 - \tfrac{1}{240}\omega^2 \cos[t\omega]y^5,
\end{aligned}
\tag{5.36}
$$

Nonlinear Partial Differential Equations are solved by the combination of homotopy perturbation method (HPM) along with Laplace Transform technique in the following examples.

1. Consider non-linear advection equation
   $u_t + uu_x = 0$, with initial condition
   $u(x,0) = -x$.
   Taking Laplace transform on both sides and applying initial condition
   $$u(x,s) = -\frac{x}{s} - \frac{1}{s}L(uu_x),$$
   Taking inverse Laplace transform
   $$u(x,t) = -x - L^{-1}\left[\frac{1}{s}L(uu_x)\right],$$

   Now applying Homotopy Perturbation method,
   $$\sum_{n=0}^{\infty} p^n u_n(x,t) = -x - p\left(L^{-1}\left[\frac{1}{s}L[\sum_{n=0}^{\infty} p^n H_n(u)]\right]\right),$$
   where $H_n(u)$ are He's polynomials representing nonlinear terms
   $H_0(u) = u_0 u_{0x}$
   $H_1(u) = u_0 u_{1x} + u_1 u_{0x}$
   $H_2(u) = u_0 u_{2x} + u_1 u_{1x} + u_2 u_{0x}$
   . . . .
   Comparing the coefficients of like powers of p

$$p^0 : u_0(x,t) = -x$$

$$p^1 : u_1(x,t) = -L^{-1}\left[\frac{1}{s}L[H_0(u)]\right] = -xt$$

$$p^2 : u_2(x,t) = -L^{-1}\left[\frac{1}{s}L[H_1(u)]\right] = -xt^2$$

Similarly

$$p^3 : u_3(x,t) = -xt^3$$
$$p^4 : u_4(x,t) = -xt^4$$

. . . .

So that the solution becomes

$$u(x,t) = u_0 + u_1 + u_2 + \cdots$$
$$u(x,t) = -x - xt - xt^2 - xt^3 - xt^4 - \cdots$$
$$u(x,t) = -x(1 + t + t^2 + t^3 + \cdots) \qquad \text{series form}$$
$$u(x,t) = -\frac{x}{(1-t)} \qquad \text{closed form}$$

is an exact solution.

2. Consider non-linear Schrodinger's equation

$iu_t + u_{xx} + 2|u|^2 u = 0$ with initial condition $u(x,0) = e^{ix}$

Applying Laplace Transform on both sides of (1) we get

$$L[u(x,t)] = \frac{e^{ix}}{s} + \frac{1}{s}iL[u_{xx} + 2u^2\bar{u}]$$

where $u^2\bar{u} = |u|^2 u$ and $\bar{u}$ is the conjugate of u.

The inverse of Laplace transform implies that

$$u(x,t) = e^{ix} + L^{-1}\{\frac{1}{s}iL[u_{xx} + 2u^2\bar{u}]\}$$

Now applying the HPM, we get

$$\sum_{n=0}^{\infty} p^n u_n(x,t) = e^{ix} + p\left(L^{-1}\left[\frac{1}{s}iL[\sum_{n=0}^{\infty} p^n u_n(x,t)_{xx} + \sum_{n=0}^{\infty} p^n H_n(u)]\right]\right),$$

Where $H_n(u)$ are He's Polynomials representing nonlinear terms. The first few components of He's polynomials are

$$H_0(u) = 2u_0^2\bar{u}_0$$
$$H_1(u) = 2(u_0^2\bar{u}_1 + 2u_1 u_0 \bar{u}_0)$$

. . . .

Comparing the coefficients of like powers of p, we have

$$p^0 : u_0(x,t) = e^{ix}$$

$$p^1 : u_1(x,t) = L^{-1}\left[\frac{1}{s}iL(u_0)_{xx} + [H_0(u)]\right] = ite^{ix}$$

$$p^2 : u_2(x,t) = L^{-1}\left[\frac{1}{s}iL(u_1)_{xx} + [H_1(u)]\right] = \frac{(it)^2}{2!}e^{ix}$$

Similarly

$$p^3 : u_3(x,t) = \frac{(it)^3}{3!}e^{ix}$$

. . . .

So that the solution becomes

$$u(x,t) = u_0 + u_1 + u_2 + \cdots = e^{ix}\left(1 + it + \frac{(it)^2}{2!} + \frac{(it)^3}{3!} + \cdots\right)$$

In the closed form,

$u(x,t) = e^{i(x+t)}$ is an exact solution.

Further Reading

For those eager to delve deeper into the subject matter explored in this book and expand their knowledge, we have compiled a list of research papers and academic resources. These references provide a more in-depth understanding of the topics covered and offer valuable insights from experts in the field. We encourage you to explore these links to gain a broader perspective and to continue your journey of discovery and learning.

1. https://www.irjet.net/archives/V8/i8/IRJET-V8I8497.pdf

2. https://ijsrcseit.com/paper/CSEIT218227.pdf

3. https://arxiv.org/abs/1804.05735

4. https://core.ac.uk/download/pdf/82685497.pdf

5. https://www.sciencedirect.com/science/article/pii/S0895717711005863/

6. http://www.tjprc.org/publishpapers/-1347974846-8-Maths%20-%20IJMCAR%20-%20Solving%20-%20G%20.%20Hariharan.pdf

Bibliography

[1] William E. Wiesel (2010). *Modern Astrodynamics*. Ohio: Aphelion Press. p. 107. ISBN 978-145378-1470.

[2] Bransden, B. H.; Joachain, C. J. (1999). *Quantum Mechanics* (2nd ed.). p. 443. ISBN 978-0582356917.

[3] Dirac, P.A.M. (1 March 1927). *"The Quantum Theory of Emission and Absorption of Radiation"*. Proceedings of the Royal Society A. 114 (767): 243–265. Bibcode:1927RSPSA.114..243D. doi:10.1098/rspa.1927.0039. JSTOR 94746. See equations (24) and (32).

[4] V.I. Arnol'd, *"Mathematical methods of classical mechanics"* , Springer (1978) (Translated from Russian)

[5] J.K. Moser, *"Lectures on Hamiltonian systems"*, Amer. Math. Soc. (1968)

[6] A.N. Kolmogorov, *"On dynamic systems with an integral invariant on a torus"* Dokl. Akad. Nauk SSSR , 93 : 5 (1953) pp. 763–766 (In Russian)

[7] N.N. [N.N. Bogolyubov] Bogoliuboff, N.M. [N.M. Krylov] Kryloff, *"Introduction to non-linear mechanics"* , Kraus (1970) (Translated from Russian)

[8] N.N. Bogolyubov, Yu.A. Mitropol'skii, *"Asymptotic methods in the theory of non-linear oscillations"* , Hindushtan Publ. Comp. , Delhi (1961) (Translated from Russian)

[9] N.N. Bogolyubov, *"On some statistical methods in mathematical physics"* , Kiev (1945) (In Russian)

[10] A.A. Dorodnitsyn, *"Asymptotic solution of the van der Pol equation"* Priklad. Mat. Mekh. , 11 (1947) pp. 313–328 (In Russian) (English abstract)

[11] A.N. Tikhonov, "*On dependence of solutions of differential systems on a small parameter*" Mat. Sb. , 22 (64) : 2 pp. 193–204 (In Russian)

[12] L.S. Pontryagin, "*Asymptotic behaviour of solutions of a system of differential equations with small parameter in front of the highest derivative*" Izv. Akad. Nauk SSSR Ser. Mat. , 21 (1957) pp. 605–626 (In Russian)

[13] E.F. Mishchenko, "*Asymptotic calculation of periodic solutions of systems of differential equations with a small parameter in front of the derivatives*" Izv. Akad. Nauk SSSR Ser. Mat. , 21 (1957) pp. 627–654 (In Russian)

[14] H. Poincar'e, *Les M´ethodes Nouvelles de la M´echanique C´eleste*, Gauthier Villars, Paris (1892)

[15] C. Delaunay, *M´emoire sur la th´eorie de la Lune*, M´em. de l'Acad. des Sciences 28 (1860) and 29 (1867)

[16] Martin C. Gutzwiller, "*Moon-Earth-Sun: The oldest three-body problem*", Rev. Mod. Phys. 70, 589 – Published 1 April 1998

[17] Cropper, William H. (2004), *Great Physicists: The Life and Times of Leading Physicists from Galileo to Hawking*, Oxford University Press, p. 34, ISBN 978-0-19-517324-6.

[18] A.N. Kolmogorov, "*On dynamic systems with an integral invariant on a torus*" Dokl. Akad. Nauk SSSR , 93 : 5 (1953) pp. 763–766 (In Russian)

[19] V.I. Arnol'd, "*Mathematical methods of classical mechanics*" , Springer (1978) (Translated from Russian)

[20] J.K. Moser, "*Lectures on Hamiltonian systems*" , Amer. Math. Soc. (1968)

[21] N.N. Bogolyubov, Yu.A. Mitropol'skii, "*Asymptotic methods in the theory of non-linear oscillations*" , Hindushtan Publ. Comp. , Delhi (1961) (Translated from Russian)

[22] C.L. Siegel, "*Vorlesungen über Himmelmechanik*" , Springer (1956)

[23] V.K. Abalakin, et al., "*Handbook of celestial mechanics and astrodynamics*" , Moscow (1971) (In Russian)

[24] H. Poincaré, "*Les méthodes nouvelles de la mécanique céleste*" , 1–3 , Gauthier-Villars (1899)

[25] C.L. Siegel, J. Moser, "*Lectures on celestial mechanics*" , Springer (1971)

[26] V.I. Arnol'd, "*Mathematical methods of classical mechanics*" ,

Springer (1978) (Translated from Russian)

[27] R. Abraham, J.E. Marsden, "*Foundations of mechanics*" , Benjamin (1978)

[28] G.N. Duboshin, "*Celestial mechanics. Analytical and qualitative methods*" , Moscow (1978) (In Russian)

[29] M.F. Subbotin, "*Introduction to theoretical astronomy*" , Moscow (1968) (In Russian)

[30] KAM theory, *cf. Quasi-periodic motion*, applies to the restricted three-body problem.

[31] H. Poincaré, "*Les méthodes nouvelles de la mécanique céleste*" , 1–3 , Gauthier-Villars (1892–1899)

[32] V.I. Arnol'd, A. Avez, "*Problèmes ergodiques de la mécanique classique*" , Gauthier-Villars (1967) (Translated from Russian)

[33] C.L. Siegel, J. Moser, "*Lectures on celestial mechanics*" , Springer (1971)

[34] R. Abraham, J.E. Marsden, "*Foundations of mechanics*" , Benjamin/Cummings (1978)

[35] WM. B. BUSH (1971), *On the Lagerstrom mathematical model for viscous flow at low Reynolds number*, this Review, to appear.

[36] R. G. CASTEN (1970), *Methodsfor deriving conservation laws*, Doctoral thesis, California Institute of Technology, Pasadena.

[37] I. E. CHANG (1961), *Navier-Stokes solutions at large distancesfrom a finite body*, J. Math. Mech., 10, pp. 811-876.

[38] D. S. COHEN AND P. A. LAGERSTROM, *Asymptotic theory of a model equation*, to be published.

[39] J. D. COLE (1968), *Perturbation Methods in Applied Mathematics*, Blaisdell, Waltham, Mass.

[40] A. ERDILYI (1956), *Asymptotic Expansions*, Dover, New York.

[41] R. P. FEYNMAN AND P. A. LAGERSTROM (1956), *Remarks on high Reynolds number flows infinite domains*, Proc. IX International Congress on Applied Mechanics, vol. 3 (Brussels, 1956).

[42] L. E. FRAZNKEL (1969), *On the method of matched asymptotic expansions*, Proc. Cambridge Philos. Soc., Part I, 65, pp. 209-231. Part II, 65, pp. 233-261; Part III, 65, pp. 263-283.

[43] K. O. FRIEDRICHS (1953), *Special Topics in Fluid Dynamics*, New York University Press, New York; see also VON MISES AND FRIEDRICHS (1953), Lecture Notes on Fluid Dynamics.

[44] S. GOLDSTEIN (1960), *Lectures on Fluid Mechanics*, Interscience, New York.

[45] SAUL KAPLUN (1954), *The role of coordinate systems in boundary-layer theory*, Z. Angew. Math. Phys., 5, pp. 111-135.

[46] —(1957), *Low Reynolds numberflow past a circular cylinder*, J. Math. Mech., 6, pp. 595-603.

[47] (1967), *Fluid Mechanics and Singular Perturbations*, P. A. Lagerstrom, L. N. Howard and C. S. Liu, eds., Academic Press, New York.

[48] SAUL KAPLUN AND P. A. LAGERSTROM (1957), *Asymptotic expansions of Navier-Stokes solutions for small Reynolds numbers*, J. Math. Mech., 6, pp. 585-593.

[49] J. KEVORKIAN (1966), *Space Mathematics, Part III*, Lectures in Applied Mathematics, vol. 7, Amer. Math. Soc., Providence, R.I.

[50] P. A. LAGERSarRO (1957), *Note on preceding two papers*, J. Math. Mech., 6, pp. 605-606.

[51] (1961), *Mthodes asymptotiques pour l'tude des Oquations de Navier-Stokes*, Lecture notes, Institut Henri Poincar6, Paris.

[52] (1964), *Laminar Flow Theory, High Speed Aerodynamics and Jet Propulsion*, F. K. Moore, ed., vol. IV, Princeton University Press, Princeton.

[53] P. A. LAGERSXROM AND J. D. COLE (1955), *Examples illustrating expansion procedures for the Navier-Stokes equations*, J. Rational Mech. Anal., 4, pp. 817-882.

[54] G. E. LAXXA (1951), *Singular perturbation problems*, Doctoral thesis, California Institute of Technology, Pasadena.

[55] HARTMtJX LEGNER (1971), *On optimal coordinates in boundary-layer theory*, Doctoral thesis, Stanford University, Stanford, Calif.

[56] L. PRANDTL (1905), *giber Flftssigkeiten bei sehr kleiner Reibung*, Verh. III Internat. Math. Kongr. (Heidelberg, 1905), Teubner, Leipzig, pp. 484-491; English transl., Tech. Memo. 452, NACA, 1928.

[57] I. PROUDMAN AND J. R. A. PARSON (1957), *Expansions at small Reynolds numbers jbr the flow past a sphere and a circular cylinder*, J. Fluid Mech., 2, pp. 237-262.

[58] M. VAN DriVE (1958), *The paraboloid of revolution in subsonic flow*, J. Math. and Phys., 37, pp. 38-51.

[59] (1962), *Higher approximations in boundary-layer theory*, J. Fluid Mech., 14, pp. 161-177 and 481-495.

[60] –(1964), *Perturbation Methods in Fluid Mechanics*, Academic Press, New York.

[61] (1964a), *Higher approximations in boundary-layer theory*. Part 3: Parabola in uniJbrm stream, J. Fluid Mech., 19, pp. 145-159.

[62] WOLFGANG WASOW (1965), *Asymptotic Expansions for Ordinary Differential Equations*, Interscience, New York.

[63] J.H. He, Homotopy perturbation technique, *Comp. Meth. Appl. Mech. Eng.* 178 (1999) 257-262.

[64] J.H. He, A coupling method of homotopy technique and a perturbation technique for non linear problems, *Int. J. Non linear Mech.* 35 (2000) 37-43.

[65] J.H. He, Homotopy perturbation method: A new nonlinear analytical technique, *Appl. Math. Comput.* 135 (2003) 73-79.

[66] J.H. He, A simple perturbation approach to Blasius equation, *Appl. Math. Comput.* 140 (2003) 217-222.

[67] J.H. He, Application of homotopy perturbation method to nonlinear wave equations, *Chaos Solitons Fractals* 26 (2005) 295-700.

[68] J.H. He, Homotopy perturbation method for solving boundary value problems, *Phys. Lett.* A 350 (2006) 87-88.

[69] A. Yildirim, Solution of BVPs for fourth-order integro-differential equations by using homotopy perturbation method, *Comput. Math. Appl.* 56 (12) (2008) 3175-3180.

[70] S. Abbasbandy, Homotopy perturbation method for quadratic Riccati differential equation and comparison with Adomian's decomposition method, *Appl. Math. Comput.* 172 (2006) 485-490.

[71] A. Golbabai, B. Keramati, Modified homotopy perturbation method for solving Fredholm integral equations, *Chaos Solitons Fractals* 37 (5) (2008) 1528-1537.

[72] M. Ghasemi, M. Tavassoli Kajani, E. Babolian, Numerical solutions of nonlinear Volterra-Fredholm integral equations by using homotopy perturbation method, *Appl. Math. Comput.* 188 (2007) 446-449.

[73] J. Biazar, H. Ghazvini, He's homotopy perturbation method for solving systems of Volterra integral equations, *Chaos Solitons Fractals* (2007) doi:10. 1016/j.chaos.2007.01.108.

[74] Z. Odibat, S. Momani, A reliable treatment of homotopy perturbation method for Klein-Gordon equations, *Phys. Lett.* A 365 (2007) 351-357.

[75] M.S.H. Chowdhury, I. Hashim, Application of homotopy perturbation method to Klein-Gordon and sine-Gordon equations, *Chaos Solitons Fractals* (2007) doi:10.1016/j.chaos.2007.06.091.

[76] M.S.H. Chowdhury, I. Hashim, Solutions of class of singular second-order IVPs by homotopy perturbation method, *Phys. Lett. A* 365 (2007) 439-447.

[77] O. Abdul Aziz, I. Hashim, S. Momani, Application of homotopy perturbation method to fractional IVPs, *J. Comput. Appl. Math.* 216 (2) (2008) 574-584.

[78] J. Biazar, H. Ghazvini, Exact solution of nonlinear Schrödinger equations by He's homotopy perturbation method, *Phys. Lett. A* 366 (2007) 79-84.

[79] M.S.H. Chowdhury, I. Hashim, Solutions of time-dependent Emden-Fowler type equations by homotopy perturbation method, *Phys. Lett. A* 368 (2007) 305-313.

[80] D.D. Ganji, H. Tari, M. Bakhshi Jooybari, Variational iteration method and homotopy perturbation method for nonlinear evolution equations, *Int. J. Comput. Math. Appl.* 54 (2007) 1018-1027.

[81] A. Yildirim, He's Homotopy Perturbation Method for nonlinear differential-difference equations, *Int. J. Comput. Math.* (2008) doi:10.1080/ 00207160802247646.

[82] A. Yildirim, The homotopy perturbation method for approximate solution of the modified KdV equation, *Z. Naturforsch.* 63a (2008) 621-626.

[83] J.H. He, Recent development of the homotopy perturbation method, *Topol. Methods. Non linear Anal.* 31 (2008) 205-209.

[84] J.H. He, An elementary introduction to the homotopy perturbation method, *Comput. Math. Appl.* 57 (2009) 410-412.

[85] S.J. Liao, An approximate solution technique not depending on small parameters: A special example, *Int. J. Non linear Mech.* 36 (1995) 371-180.

[86] S.J. Liao, Beyond Perturbation: Introduction to Homotopy Analysis Method, *CRC, Boca Raton,* 2004.

[87] J. Biazar, H. Ghazvini, Convergence of the homotopy perturbation method for partial differential equations, *Non Linear Anal. RWA* (2008) doi:10.1016/ j.nonrwa.2008.07.002.

[88] A. Sami Bataineh, M.S.M. Noorani, I. Hashim, The homotopy analysis method for Cauchy reaction-diffusion problems, *Phys. Lett.* A 372

(5) (2008) 613-618.

[89] A. Sami Bataineh, M.S.M. Noorani, I. Hashim, Solutions of time-dependent Emden-Fowler type equations by homotopy analysis method, *Phys. Lett. A* 371 (2007) 72-82.

[90] A. Yildirim, T. Özi, Solutions of singular IVPs of Lane-Emden type by homotopy perturbation method, *Phys. Lett. A* 369 (2007) 70-76.

[91] Z. Odibat, A new modification of the homotopy perturbation method for linear and nonlinear operators, *Appl. Math. Comput.* 189 (2007) 746-753

[92] J. H. He, Homotopy perturbation technique, *Comput. Methods Appl. Mech. Engrg.*, 178 (3-4) (1999), 257-262.

[93] J. H. He, A coupling method of homotopy technique and perturbation technique for nonlinear problems, *Int. J. Nonlinear Mech.*, 35 (1) (2000), 37-43.

[94] J. H. He, Some asymptotic method for strongly nonlinear equations, *Internat. J. Modern Phys. B*, 20 (2006), 1141-1199.

[95] J. H. He, A new perturbation technique which is also valid for large parameters, *J. Sound Vibration*, 229 (2000), 1257-1263.

[96] M. El. Shahed, Application of He's Homotopy perturbation method to Volterra's integro differential equation, *Int. J. Nonlinear Sci. Numer. Simul.*, 6 (2) (2005), 163-168.

[97] J. H. He, The homotopy perturbation method for nonlinear oscillators with discontinuities, *Appl. Math. Comput.*, 151 (2004), 287-292.

[98] J. H. He, Homotopy perturbation method for bifurcation of nonlinear problems, *Int. J. Nonlinear Sci. Numer. Simul.*, 6 (2) (2005), 207-208.

[99] J. H. He, Periodic solutions and bifurcations of delay differential equations, *Phys. Lett. A*, 374(4-6)(2005)228-230.

[100] J. H. He, Application of homotopy perturbation method to nonlinear wave equations, *Chaos Solitons Fractals*, 26 (3) (2005), 695-700.

[101] J. H. He, Homotopy perturbation method for solving boundary value problems, *Phys. Lett. A*, 350 (1-2) (2006), 87-88. J. H. He, Homotopy perturbation method: a new nonlinear analytic technique, *Appl. Math. Comput.*, 135 (2003), 73-79.

[102] J. H. He, Comparison of homotopy perturbation method and homotopy analysis method, *Appl. Math. Comput.*, 156 (2004), 527-539.

[103] G. Adomian, Frontier Problem of Physics: The Decomposition Method, *Kluwer Academic Publishers*, Boston, 1994.

[104] V.D. Gejji, H. Jafari, Adomian decomposition: a tool for solving a system of fractional differential equations, *J. Math. Anal. Appl.* 301 (2005) 508–518.

[105] H. Jafari, V.D. Gejji, Revised Adomian decomposition method for solving a system of nonlinear equations, *Appl. Math. Comput.* 175 (2006) 1–7.

[106] C. Chun, H. Jafari, Y. Kim, Numerical method for the wave and nonlinear diffusion equations with the homotopy perturbation method, *Comput. Math. Appl.* 57 (2009) 1226–1231.

[107] A.M. Wazwaz, A study of boundary-layer equation arising in an incompressible fluid, *Appl. Math. Comput.* 87 (1997) 199–204.

[108] S.J. Liao, An optimal homotopy-analysis approach for strongly non-linear differential equations, *Commun. Nonlinear Sci. Numer. Simul.* 15 (2010) 2003–2016.

[109] S.J. Liao, On the relationship between the homotopy analysis method and Euler transform, *Commun. Nonlinear Sci. Numer. Simul.* 15 (2010) 1421–1431.

[110] J. Cheng, S.P. Zhu, S.J. Liao, An explicit series approximation to the optimal exercise boundary of American put options, *Commun. Nonlinear Sci. Numer. Simul.* 15 (2010) 1148–1158.

[111] S. Nadeem, A. Hussain, M. Khan, HAM solutions for boundary layer flow in the region of the stagnation point towards a stretching sheet, *Commun. Nonlinear Sci. Numer. Simul.* 15 (2010) 457–481.

[112] S. Nadeem, A. Hussain, M. Khan, Stagnation flow of a Jeffrey fluid over a shrinking sheet, *Z. Nat. Forsch. A 65a* (2010) 1–9.

[113] A. Yildirim, T. Ozis, Solutions of singular IVPs of Lane–Emden type by homotopy perturbation method, *Phys. Lett.* A 369 (2007) 70–76.

[114] A. Yildirim, Exact solutions of nonlinear differential–difference equations by He's homotopy perturbation method, *Int. J. Nonlinear Sci. Numer. Simul.* 9 (2008) 111–114.

[115] A. Yildirim, Variational iteration method for modified Ca-massa–Holm and Degasperis–Procesi equations, *Int. J. Numer. Methods Biomed. Eng.* 26 (2010) 266–272.

[116] A. Yildirim, Exact solutions of Poisson equation for electrostatic potential problems by means of the variational iteration method, *Int. J. Nonlinear Sci. Numer. Simul.* 10 (2009) 867–871.

[117] A. Yildirim, T. Ozis, Solutions of singular IVPs of Lane–Emden

type by the variational iteration method, *Nonlinear Anal. Theory Methods Appl.* 70 (2009) 2480–2484.

[118] A.M. Wazwaz, The variational iterative method for solving two forms of Blasius equation on a half-infinite domain, *Appl. Math. Comput.* 188 (2007) 485–491.

[119] M. Khan, M. Hussain, Application of Laplace decomposition method on semi infinite domain, *Numer. Algorithms* 56 (2011) 211–218.

[120] M. Khan, M.A. Gondal, A new analytical solution of foam drainage equation by Laplace decomposition method, *J. Adv. Res. Differ. Eqs.* 2 (2010) 53–64.

[121] M. Khan, M.A. Gondal, Restrictions and improvements of Laplace decomposition method, *J. Adv. Res. Sci. Comput.* 3 (2011) 8–14.

[122] M. Hussain, M. Khan, Modified Laplace decomposition method, *Appl. Math. Sci.* 4 (2010) 1769–1783.

[123] M.A. Gondal, M. Khan, Homotopy perturbation method for nonlinear exponential boundary layer equation using Laplace transformation, He's polynomials and Pade technology He's polynomials and Pade technology, *Int. J. Nonlinear Sci. Numer. Simul.* 11 (2010) 1145–1153.

[124] J.P. Boyd, Padé approximant algorithm for solving nonlinear ordinary differential equation boundary value problems on an unbounded domain, *Computer Physics* 11 (1997) 299–303.

[125] G.A. Baker, Essentials of Padé Approximants, *Academic Press, London,* 1975

[126] M. Fathizadeh, F. Rashidi, Boundary layer convective heat transfer with pressure gradient using Homotopy Perturbation Method (HPM) over a flat plate, *Chaos Solitons and Fractals* 42 (2009) 2413–2419.

[127] J-Huan He, Homotopy perturbation technique, *Computer methods in applied mechanics and engineering.* 178 (1999) 257–262.

[128] J-Huan He, A coupling method of homotopy technique and a perturbation technique for nonlinear problems, *Int. J. Non linear Mech.* 35 (2000) 37–43.

[129] Bongsoo Jang, Exact solutions to one dimensional non-homogeneous parabolic problem by the homogeneous Adomian decomposition method, *Applied Mathematics and Computation* 186 (2007) 969–979.

[130] A. Barari, Abdoul R.Ghotbi, T. Barari and D.D.Ganji, Applica-

tion of He's Homotopy Perturbation Method to Non-Homogeneous Parabolic Problems, *International Journal of Applied Mathematics and Computation* 1 (2009) 59–66.

[131] Abdoul R. Ghotbi, A. Barari, M. Omidvar, D.D. Ganji, Application of Variational Iteration Method to Parabolic Problems, *Applied Mathematical Sciences* 3 (2009) 927–934.

[132] A.M. Lyapunov, The General Problem of the Stability of Motion, Taylor & Francis, London, UK, 1992, English translation.

[133] J.H. He, Homotopy perturbation technique, Computer Methods in Applied Mechanics and Engineering 178 (1999) 257–262.

[134] N.H. Sweilam, M.M. Khader, Exact solutions of some coupled non-linear partial differential equations using the homotopy perturbation method, Computers & Mathematics with Applications 58 (2009) 2134–2141.

[135] J. Saberi-Nadjafi, A. Ghorbani, He's homotopy perturbation method: an effective tool for solving nonlinear integral and integro-differential equations, Computers & Mathematics with Applications 58 (2009) 1345–1351.

[136] A.V. Karmishin, A.I. Zhukov, V.G. Kolosov, Methods of Dynamics Calculation and Testing for Thin-Walled Structures, Mashinostroye-nie, Moscow, Russia, 1990.

[137] R. Hirota, Exact solutions of the Korteweg–de Vries equation for multiple collisions of solitons, Physical Review Letters 27 (1971) 1192–1194.

[138] A.M. Wazwaz, On multiple soliton solutions for coupled KdV–mkdV equation, Nonlinear Science Letters A 1 (2010) 289–296.

[139] G. Adomian, Solving Frontier Problems of Physics: The Decomposition Method, Kluwer Acad. Publ., Boston, 1994.

[140] G.C. Wu, J.H. He, Fractional calculus of variations in fractal space-time, Nonlinear Science Letters A 1 (2010) 281–287.

[141] J.H. He, Variational iteration method—a kind of nonlinear analytical technique: some examples, International Journal of Nonlinear Mechanics 34 (1999) 699–708.

[142] J.H. He, X.H. Wu, Variational iteration method: new development and applications, Computers & Mathematics with Applications 54 (2007) 881–894.

[143] J.H. He, G.C. Wu, F. Austin, The variational iteration method which should be followed, Nonlinear Science Letters A 1 (2009) 1–30.

[144] L.A. Soltani, A. Shirzadi, A new modification of the variational iteration method, Computers & Mathematics with Applications 59 (2010) 2528–2535.

[145] N. Faraz, Y. Khan, A. Yildirim, Analytical approach to two-dimensional viscous flow with a shrinking sheet via variational iteration algorithm-II, Journal of King Saud University (2010) doi:10.1016/j.jksus.2010.06.010.

[146] G.C. Wu, E.W.M. Lee, Fractional variational iteration method and its application, Physics Letters A (2010) doi:10.1016/j.physleta.2010.04.034.

[147] E. Hesameddini, H. Latifizadeh, Reconstruction of variational iteration algorithms using the Laplace transform, International Journal of Nonlinear Sciences and Numerical Simulation 10 (2009) 1377–1382.

[148] C. Chun, Fourier-series-based variational iteration method for a reliable treatment of heat equations with variable coefficients, International Journal of Nonlinear Sciences and Numerical Simulation 10 (2009) 1383–1388.

[149] G. Adomian, Solution of physical problems by decomposition, Computers & Mathematics with Applications 2 (1994) 145–154.

[150] A.M. Wazwaz, A comparison between the variational iteration method and adomian decomposition method, Journal of Computational and Applied Mathematics 207 (2007) 129–136.

[151] M.A. Abdou, A.A. Soliman, New applications of variational iteration method, Physica D: Nonlinear Phenomena 211 (2005) 1–8.

[152] M. Dehghan, Weighted finite difference techniques for the one-dimensional advection–diffusion equation, Applied Mathematics and Computation 147 (2004) 307–319.

[153] D.D. Ganji, A. Sadighi, Application of He's homotopy perturbation method to nonlinear coupled systems of reaction diffusion equations, International Journal of Nonlinear Sciences and Numerical Simulation 7 (2006) 411–418.

[154] Y. Khan, F. Austin, Application of the Laplace decomposition method to nonlinear homogeneous and non-homogenous advection equations, Zeitschrift fuer Naturforschung 65a (2010) 1–5.

[155] S.T. Mohyud-Din, A. Yildirim, Homotopy perturbation method for advection problems, Nonlinear Science Letters A 1 (2010) 307–312.

[156] J.H. He, Homotopy perturbation method: a new nonlinear analytical technique, Applied Mathematics and Computation 135 (2003)

73–79.

[157] J.H. He, Comparison of homotopy perturbation method and homotopy analysis method, Applied Mathematics and Computation 156 (2004) 527–539.

[158] J.H. He, The homotopy perturbation method for nonlinear oscillators with discontinuities, Applied Mathematics and Computation 151 (2004) 287–292.

[159] J.H. He, Homotopy perturbation method for bifurcation of nonlinear problems, International Journal of Nonlinear Sciences and Numerical Simulation 6 (2005) 207–208.

[160] J.H. He, Some asymptotic methods for strongly nonlinear equation, International Journal of Modern Physics 20 (2006) 1144–1199.

[161] J.H. He, Homotopy perturbation method for solving boundary value problems, Physics Letters A 350 (2006) 87–88.

[162] M. Rafei, D.D. Ganji, Explicit solutions of helmhotz equation and fifth-order KdV equation using homotopy perturbation method, International Journal of Nonlinear Sciences and Numerical Simulation 7 (2006) 321–328.

[163] A.M. Siddiqui, R. Mahmood, Q.K. Ghori, Thin film flow of a third grade fluid on a moving belt by He's homotopy perturbation method, International Journal of Nonlinear Sciences and Numerical Simulation 7 (2006) 7–14.

[164] D.D. Ganji, The applications of He's homotopy perturbation method to nonlinear equation arising in heat transfer, Physics Letters A 335 (2006) 337–3341.

[165] L. Xu, He's homotopy perturbation method for a boundary layer equation in unbounded domain, Computers & Mathematics with Applications 54 (2007) 1067–1070.

[166] J.H. He, An elementary introduction of recently developed asymptotic methods and nanomechanics in textile engineering, International Journal of Modern Physics 22 (2008) 3487–4578.

[167] J.H. He, Recent developments of the homotopy perturbation method, Topological Methods in Nonlinear Analysis 31 (2008) 205–209.

[168] E. Hesameddini, H. Latifizadeh, An optimal choice of initial solutions in the homotopy perturbation method, International Journal of Nonlinear Sciences and Numerical Simulation 10 (2009) 1389–1398.

[169] E. Hesameddini, H. Latifizadeh, A new vision of the He's homotopy

perturbation method, International Journal of Nonlinear Sciences and Numerical Simulation 10 (2009) 1415–1424.

[170] J. Biazar, M. Gholami Porshokuhi, B. Ghanbari, Extracting a general iterative method from an adomian decomposition method and comparing it to the variational iteration method, Computers & Mathematics with Applications 59 (2010) 622–628.

[171] S.A. Khuri, A Laplace decomposition algorithm applied to a class of nonlinear differential equations, Journal of Applied Mathematics 1 (2001) 141–155.

[172] E. Yusufoglu, Numerical solution of Duffing equation by the Laplace decomposition algorithm, Applied Mathematics and Computation 177 (2006) 572–580.

[173] Yasir Khan, An effective modification of the Laplace decomposition method for nonlinear equations, International Journal of Nonlinear Sciences and Numerical Simulation 10 (2009) 1373–1376.

[174] Yasir Khan, Naeem Faraz, A new approach to differential difference equations, Journal of Advanced Research in Differential Equations 2 (2010) 1–12.

[175] S. Islam, Y. Khan, N. Faraz, F. Austin, Numerical solution of logistic differential equations by using the Laplace decomposition method, World Applied Sciences Journal 8 (2010) 1100–1105.

[176] M. Madani, M. Fathizadeh, Homotopy perturbation algorithm using Laplace transformation, Nonlinear Science Letters A 1 (2010) 263–267.

[177] M.A. Noor, S.T. Mohyud-Din, Variational homotopy perturbation method for solving higher dimensional initial boundary value problems, Mathematical Problems in Engineering 2008 (2008) 11. Article ID 696734, doi:10.1155/2008/696734.

[178] A. Ghorbani, J. Saberi-Nadjafi, He's homotopy perturbation method for calculating adomian polynomials, International Journal of Nonlinear Sciences and Numerical Simulation 8 (2007) 229–232.

[179] A. Ghorbani, Beyond adomian's polynomials: He polynomials, Chaos Solitons Fractals 39 (2009) 1486–1492.

[180] S.T. Mohyud-Din, M.A. Noor, K.I. Noor, Traveling wave solutions of seventh-order generalized KdV equation using He's polynomials, International Journal of Nonlinear Sciences and Numerical Simulation 10 (2009) 227–233.

[181] Yasir Khan,Qingbiao Wu,Homotopy perturbation transform

method for nonlinear equations using He's polynomials, Computers and Mathematics with Applications, 61 (8) (2011)1963–1967.

[182] Majid Khan,Muhammad Asif Gondal, Iqtadar Hussain,S. Karimi Vanani,A new comparative study between homotopy analysis transform method and homotopy perturbation transform method on a semi infinite domain, Mathematical and computer modeling, 55(3-4) (2012)1143-1150.

[183] V.G.Gupta,Sumit Gupta,Application of Homotopy perturbation transform method for solving heatlike and wave-like equations with variable coefficients,International Journal of Mathematical Archive,2(9) (2011).

[184] Majid Khan, Muhammad Asif Gondal and Sunil Kumar,A novel Homotopy perturbation algorithm for linear and nonlinear system of partial differential equations, World Applied Sciences journal 12(12),(2011) 2352-2357.

[185] G.Adomain,Frontier problem of physics: the decomposition method, Boston:KLuwer Academic publishers;1994.

[186] Abdul-Masjid wazwaz,The decomposition method applied to system of partial differential equations and to reaction diffusion Brusselator model, Applied mathematics and computation 110(2000)251-264.

[187] Mohammed Ghoreishi, Adomain Decomposition Method for Nonlinear wave-like equations with variable coefficient,Applied Mathematical Sciences,4(49) (2010) 2431-2444.

[188] J.H. He, X.Wu,Variational iteration method:new development and applications, Computer and Mathematics with Applications 54(2007)881-894.

[189] J.H.He, G.C.Wu, F.Austin, The variational iteration method which should be followed,Nonlinear Science Letter A (2009)1-30.

[190] L.A.Soltani, A.Shirzadi, A new modification of variational iteration method, Computer and Mathematics with Applications 59(2010)2528-2535.

[191] J.H.He,Homotopy perturbation method:a new nonlinear analytical technique, Applied Mathematics and Computation135(2003)73-79.

[192] J.H.He,Homotopy perturbation method for boundary value problems ,Physics Letters A,350(2006),87-88.

[193] Jafar Biazar,Hossein Aminikhah,Study of convergence of Homotopy perturbation method for systems of partial differential equations,

computers and mathematics with applications 58(2009)2221-2230.

[194] Behrouz Raftari,Application of He's Homotopy perturbation method and variational iteration method for Nonlinear partial Integro-differential equations,World Applied sciences 7(4):399-404,2009.

[195] Khatereh TABATABAEI,'et.al',The differential transform method for solving heat-like and wavelikeequations with variable coefficients,Tuek J Phys,36(2012),87-98.

[196] Jafar Biazar,Fatemeh Mohammadi,Application of Differential Transform Method to the Sine-GordonEquation, International Journal of Nonlinear Science Vol.10(2010)No.2,pp190-195.

[197] Y.Keskin,G.Oturanc,Reduced differential transform method for solving Linear and Nonlinear wave equations,Iranian journal of Science & Technology,Transaction A,vol.34,No.A2, 2010.

[198] J. H. He, Homotopy perturbation technique, computer methods in Applied Mechanics and Engineering 178 (1999) 257 - 262.

[199] Lan Xu He's Homotopy perturbation method for a boundary layer equation in unbounded domain, International Journal of Computer and Mathematics application.Elsevier-54(2007)10671070.

[200] Y. Khan, Wu Q, Homotopy perturbation transform method for non-linear equations using He's polynomials, Journal of Computers and Mathematics with application, Elserier, 2010.

[201] A Ghorbani, Beyond Adomian's polynomials : He polynomials chaos solutions Fractals 39(2009)1486 − 1492.

[202] Devendra Kumar, Jagder Singh, Sushila, Application of Homotopy Perturbation Transform Method to Linear and Nonlinear Schro dinger Equations,Internation Journal of Nonlinear science Vol.16 (2013) No.3 PP 203-209.

[203] Gupta V. G. Gupta S., Homotopy perturbation Transform method for solving initial boundary value problem of variable co-efficient; International journal of non-linear science, Vol.12 (2011) No.3 PP270-277.

[204] Mohammad Madani Mahdi Fathizadeh, Yasir Khan, Ahmet Yildirim, Mathematical and Computer modelling 53 (2011), Elsevier, 1937-1945.

[205] Mansi K. Desai, S. S. Patel, A Solution of Fluid flow through Porous medium Equation by Homotopy Perturbation Transform Method, International research Journal of Engineering and Technology (IRJET)

Volume : 03, Issue : 07/July 2016.

[206] Mubashra Saleem,Aqsa Mumtaz,Tahira Amir, Qazi Muhmmad ul Hassan,Kamran Ayub, Farhana Kanwal-Homotopy Perturbation Method Using Hes Polynomial for Solving Nonlinear Differential Equations Open Science Journal of Mathematics and applications2017;5(2):8-11 Homotopy perturbation method using He's polynomial for solving non linear differential equations.

[207] Sari M, solution of the porous media equation by a compact finite differential method, mathematical problem in Engineering, Hindwi Publishing corporation, 2009.

[208] N. H. Sweilam, M. M. Khader, Exact solution of some coupled non-linear partial differential equations using the homotopy perturbation method, computers and mathematics with Application 58 (2009) 2134 - 2141. Topological methods in non-linear Analysis $31(2008)205 - 209$.

[209] Asghar Grorbani, Beyond Adomian's polynomials: He polynomials chaos solutions Fractals 39(2009) 14861492.

[210] Jagdev Singh, Devendra Kumar and Sushila Rathore, Application of Homotopy perturbation Transform method for solving linear and non-linear Klein-Gordon equations, Journal of Information and Computing Science, Vol.7, No.2, 2012, pp.131-139.

[211] J.H. He, Homotopy Perturbation technique, Computer methods in Applied Mechanics and Engineering 178(1999) 257-262.

[212] Manasi K. Desai, S.S. Patel, A solution of fluid flow through porous medium equation by Homotopy Perturbation Transform Method, International Research Journal of Engineering and Technology, Vol 03, issue 07/ July 2016.

[213] Mohamed Elbadri, Comparison between the Homotopy Perturbation method and Homotopy Perturbation Transform Method, Scientific Research PublishingApplied Mathematics, 2018, 9, 130-137.

[214] Mohamed Jleli, Sunil Kumar, Ranbir Kumar, Bessem Samet, Analytical approach for time fractional wave equations in the sense of Yang-Abdel-Aty-Cattani via the homotopy perturbation Transform Method, Alexandria Engineering Journal (2020)59, 2859-2863.

[215] Mubashra Saleem, Aqsa Mumtaz, Tahira Amir, Qazi Muhmmad ul Hassan, Kamran Ayub, Farhana KanwalHomotopy Perturbation Method Using He's Polynomial for Solving Non- linear Differential Equations, Open Science Journal of Mathematics and applications2017;5(2):8-11.

[216] N. H. Sweilam, M. M. Khader, Exact solution of some coupled non-

linear partial differential equations using the homotopy perturbation method, computers and mathematics with Application 58 (2009) 2134 - 2141. Topological methods in non-linear Analysis 31(2008) 205-209.

[217] Y. Khan, Wu Q, Homotopy perturbation transform method for non-linear equations using He 's polynomials, Journal of Computers and Mathematics with application, Elsevier, 2010.

[218] U. Filobello-Nino, H. Vazquez -Leal, Y. Khan, A. PerezSesma, A. Diaz-Sanchez, V. M. Jimenez-Fernandez, A. Herrera-May, D. Pereyra-Diaz, J. Sanchez-Orea, Laplace transform homotopy perturbation method as a powerful tool to solve nonlinear problems with boundary conditions defined on finite intervals, Comp. Appl. Math. DOI 10.1007/s40314-013-0073-z.

[219] Peter O' Neil, Advanced Engineering Mathematics, Thomson Publication, 5th Edition.

[220] Erwin Kreyszig, Advanced Engineering Mathematics, John Wiley, 10th Edition.

[221] Valery Serov, Fourier series, Fourier Transforms and their applications to Mathematical Physics, Springer, Volume 197.

[222] G. James, Advanced Modern Engineering Mathematics (3rd Edition), Pearson Education Limited, New York 1999.

[223] Lawrence C. Evans, Differential Equations, AMS 1998.

[224] http://Quora.com

[225] I. Podlubny, Fractional differential equations, vol. 198 of Mathematics in Science and Engineering, Academic Press, San Diego, Calif, USA, 1999.

[226] K. S. Miller and B. Ross, An Introduction to the Fractional Calculus and Fractional Differential Equations, John Wiley & Sons, New York, NY, USA, 2003.

[227] K. B. Oldham and J. Spanier, The Fractional Calculus, Academic Press, New York, NY, USA, 1974.

[228] K. Diethelm and N. J. Ford, "Analysis of fractional differential equations," Journal of Mathematical Analysis and Applications, vol. 265, no. 2, pp. 229-248, 2002.

[229] S. G. Samko, A. A. Kilbas, and O. I. Marichev, Fractional Integrals and Derivatives, Gordon and Breach, Yverdon, Switzerland, 1993.

[230] B. J. West, M. Bologna, and P. Grigolini, Physics of Fractal Operators, Institute for Nonlinear Science, Springer, New York, NY, USA, 2003.

[231] S. S. Ray, "Analytical solution for the space fractional diffusion equation by two-step Adomian decomposition method," Communications in Nonlinear Science and Numerical Simulation, vol. 14, no. 4, pp. 1295-1306, 2009.

[232] K. Abbaoui and Y. Cherruault, "New ideas for proving convergence of Adomian Decomposition Methods," Computers and Mathematics with Applications, vol. 32, pp. 103-108, 1995.

[233] S. Momani and Z. Odibat, "Analytical solution of a timefractional Navier-Stokes equation by Adomian decomposition method," Applied Mathematics and Computation, vol. 177, no. 2, pp. 488-494, 2006.

[234] H. Jafari and V. Daftardar-Gejji, "Solving linear and nonlinear fractional diffusion and wave equations by Adomian decomposition," Applied Mathematics and Computation, vol. 180, no. 2, pp. 488-497, 2006. [11] M. Ganjiani, "Solution of nonlinear fractional differential equations using homotopy analysis method," Applied Mathematical Modelling, vol. 34, no. 6, pp. 1634-1641, 2010.

[235] H. Jafari, C. M. Khalique, and M. Nazari, "Application of the Laplace decomposition method for solving linear and nonlinear fractional diffusion-wave equations," Applied Mathematics Letters, vol. 24, no. 11, pp. 1799-1805, 2011.

[236] P. K. Gupta and M. Singh, "Homotopy perturbation method for fractional Fornberg-Whitham equation," Computers & Mathematics with Applications, vol. 61, no. 2, pp. 250-254, 2011.

[237] S. Momani and Z. Odibat, "Homotopy perturbation method for nonlinear partial differential equations of fractional order," Physics Letters. A, vol. 365, no. 5-6, pp. 345-350, 2007.

[238] Y.-Z. Zhang, A.-M. Yang, and Y. Long, "Initial boundary value problem for fractal heat equation in the semi-infinite region by yang-laplace transform," Thermal Science, vol. 18, no. 2, pp. 677681, 2014.

[239] X.-J. Yang and D. Baleanu, "Fractal heat conduction problem solved by local fractional variation iteration method," Thermal Science, vol. 17, no. 2, pp. 625-628, 2013.

[240] X.-J. Yang, D. Baleanu, Y. Khan, and S. T. Mohyud-Din, "Local fractional variational iteration method for diffusion and wave equations on Cantor sets," Romanian Journal of Physics, vol. 59, no. 1-2, pp. 36-48, 2014.

[241] X. J. Yang, H. M. Srivastava, J. H. He, and D. Baleanu, "Cantortype cylindrical-coordinate fractional derivatives," Proceedings of the Romanian Academy, Series A, vol. 14, pp. 127-133, 2013.

[242] S. Kumar, D. Kumar, S. Abbasbandy, and M. M. Rashidi, "Analytical solution of fractional Navier-Stokes equation by using modified Laplace decomposition method," Ain Shams Engineering Journal, vol. 5, no. 2, pp. 569-574, 2014.

[243] B. Ghazanfari and A. G. Ghazanfari, "Solving fractional nonlinear Schrodinger equation by fractional complex transform method," International Journal of Mathematical Modelling o Computations, vol. 2, no. 4, pp. 277-281, 2012.

[244] K. M. Furati and N.-E. Tatar, "Longtime behavior for a nonlinear fractional model," Journal of Mathematical Analysis and Applications, vol. 332, no. 1, pp. 441-454, 2007.

[245] K. M. Furati and N.-E. Tatar, "Behavior of solutions for a weighted Cauchy type fractional problem," Journal of Fractional Calculus, vol. 28, pp. 23-42, 2005.

[246] K. M. Furati and N.-e. Tatar, "An existence result for a nonlocal fractional differential problem," Journal of Fractional Calculus, vol. 26, pp. 43-51, 2004.

[247] J.-H. He, "Homotopy perturbation technique," Computer Methods in Applied Mechanics and Engineering, vol. 178, no. 3-4, pp. 257-262, 1999.

[248] J. H. He, "Recent development of the homotopy perturbation method," Topological Methods in Nonlinear Analysis, vol. 31, pp. 205-209, 2008.

[249] J.-H. He, "The homotopy perturbation method for nonlinear oscillators with discontinuities," Applied Mathematics and Computation, vol. 151, no. 1, pp. 287-292, 2004.

[250] Z. H. Khan and W. A. Khan, "N-transform properties and applications," NUST Journal of Engineering Sciences, vol. 1, pp. 127133, 2008.

[251] F. B. M. Belgacem and R. Silambarasan, "Theory of the natural transform," Mathematics in Engineering, Science and Aerospace (MESA) Journal, vol. 3, no. 1, pp. 99-124, 2012.

[252] F. B. M. Belgacem and R. Silambarasan, "Advances in the natural transform," in Proceedings of the 9th International Conference on Mathematical Problems in Engineering, Aerospace and Sciences (ICNPAA '12), vol. 1493 of AIP Conference Proceedings, pp. 106110, Vienna, Austria, July 2012.

[253] R. Murray, Spiegel. Theory and Problems of Laplace Transform, Schaum's Outline Series, McGraw-Hill, New York, NY, USA, 1965.

[254] F. B. Belgacem and A. Karaballi, "Sumudu transform fundamental properties investigations and applications," Journal of Applied Mathematics and Stochastic Analysis, vol. 2006, Article ID 91083, 23 pages, 2006.

[255] G. K. Watugala, "Sumudu transform-a new integral transform to solve differential equations and control engineering problems," Mathematical Engineering in Industry, vol. 6, no. 4, pp. 319-329, 1998.

[256] R. Silambarasan and F. B. M. Belgacem, "Application of the natural transform to Maxwell's equations," in Proceedings of the Progress in Electromagnetics Research Symposium Proceedings (PIERS '11), pp. 899-902, Suzhou, China, September 2011.

[257] M. Rawashdeh and S. Maitama, "Finding exact solutions of nonlinear PDEs using the natural decomposition method," Mathematical Methods in the Applied Sciences, 2016.

[258] S. Maitama and S. M. Kurawa, "An efficient technique for solving gas dynamics equation using the natural decomposition method," International Mathematical Forum, vol. 9, no. 24, pp. 1177-1190, 2014.

[259] S. Maitama, "Exact solution of equation governing the unsteady flow of a polytropic gas using the natural decomposition method," Applied Mathematical Sciences, vol. 8, no. 77, pp. 38093823,2014

[260] M. S. Rawashdeh and S. Maitama, "Solving PDEs using the natural decomposition method," Nonlinear Studies, vol. 23, no. 1, pp. 63-72, 2016.

[261] M. S. Rawashdeh and S. Maitama, "Solving nonlinear ordinary differential equations using the NDM," The Journal of Applied Analysis and Computation, vol. 5, no. 1, pp. 77-88, 2015.

[262] M. S. Rawashdeh and S. Maitama, "Solving coupled system of nonlinear PDE's using the natural decomposition method," International Journal of Pure and Applied Mathematics, vol. 92, no. 5, pp. 757-776, 2014.

[263] Y. Luchko and R. Gorenflo, "An operational method for solving fractional differential equations with the Caputo derivatives," Acta Mathematica Vietnamica, vol. 24, no. 2, pp. 207-233, 1999.

[264] Ramaa Sandu and Dr. B.B. Waphare, "Review on Homotopy Perturbation Transform Method: A Powerful Tool for the Analytical Solution of Linear and Non- Linear Partial Differential Equations"

[265] Snehal Yelai, Ramaa Sandu, Vaishali M. Joshi,"An Overview of Various Analytical Methods for Solving One Dimensional Wave Equation"

Notes

Notes

Notes

Notes